In spite of the limits of our aims and achievements, we hope to have been able to give a hint of the delicate tunes evolution has played to produce the present forms of life and how difficult it would be to predict the outcome of a major perturbation. Thus we dedicate this book to all who endeavor by any means to prevent nuclear wars from occurring on our planet or its outer space.

INTRODUCTORY BIOPHYSICS

M Cerdonio
R W Noble

World Scientific

Published by

World Scientific Publishing Co Pte Ltd.
P. O. Box 128, Farrer Road, Singapore 9128.

ISBN 9971-966-33-6

Printed in Singapore by Fu Loong Lithographer Pte Ltd.

PREFACE

As far as we can discern at present, the occurrence of life on this planet is a unique event within the universe. In recent years there have been spectacular advances in both molecular biology and the physics of irreversible processes. Together these offer substantial hope that we shall be able to understand living systems in terms of the known physical laws, and thus see life as one of the many phenomena displayed by the universe in its evolution. This book is an attempt to introduce physicists and physically-oriented students of the biological sciences to this view. An introductory discussion of the definition of "living" is followed by an overview of the properties of living systems as we know them. Then selected topics, chosen because of their fundamental importance to our understanding of living systems, are presented in greater detail. Therefore this book is not a complete text of biophysical or biochemical topics. The subjects chosen for discussion are related to the origin of life, the physical requirements for ordered living systems, and the physical and chemical bases for the most fundamental phenomena displayed by living systems such as photosynthesis, energy transfer and storage, and reproduction. It is hoped that this will stimulate the interest and furnish the knowledge necessary to further explore these topics in the current literature.

CONTENTS

CONTENTS

CHAPTER 1

INTRODUCTION

Life is but day at most, sprung from night, in darkness lost.

Robert Burns

It is generally expected that a volume such as this will begin with a discussion of the definition of "life". One could argue that we should avoid such a discussion, because one may ask to what extent we actually need a definition of what life is. Nevertheless, some observations are in order.

One may try to recognize general features of all living systems, which, if needed, could allow one to go beyond the living structures based on the self-replication of nucleic acids. A living system would be defined as a system which i) instructs its own replication ii) in this process is subject to mutations and iii) competes with other systems for survival. This would allow us to recognize, as such, living creatures which may have different biochemistry than ours, if ever we have such an encounter. It is likely to include computers of the future, as science fiction likes to depict them. Besides there are other systems which may easily fall within the framework of such a definition. A secret society, for instance, would instruct self-replications of its units in order to disseminate to more and more cities. It could mutate by adjusting its internal rules either accidentally because of the turnover of the individuals which rule it or in response to the stimuli from the environment in which it operates and may compete with other secret societies for adepts. The kinetic equations to describe these phenomena would be quite similar in all these cases. This approach has the problem of giving the uncomfortable feeling that one is attempting to give an answer before the question has been thoroughly formulated. In the past this approach sometimes led to the assumption that there is some secret, special state of matter or some unique interaction crucial to living systems, which is yet to be discovered. This may be so and there is little doubt that the discussion of such a possibility has been a great driving force for past research. However, until now no hint of anything like this has been found.

Another possibility is to regard life as a physical quantity, like for instance temperature, and to seek an operative definition which will permit one to measure "how much living" a creature or object is. This approach may seem more acceptable in its methodology to physicists and may offer some promise. In Table 1 we give a simple-minded list of living systems, specifying for each, the characteristic living features they display, such as replication, motility, etc. It can be seen that there is a virtually continuous scale of complexity. However, it would be very difficult to impose clear units on this scale. We should emphasize at this point that the living systems in the list do not represent some sort of evolutionary sequence from less to more evolved. On the contrary they represent, each in its own right, very highly adapted systems, the very best of each kind, since they all survive at the same time in mutual interaction and competition in the same integrated ecological system. However, it is obviously striking how, in some sense, a paramecium is much more "living" than a virus and less than an ape. Moreover it would still appear to be an open question whether or not we want to call viroids, plasmids and viruses "living". While the last consideration motivates this discussion, we still encounter great difficulty in trying to go beyond the stage outlined in Table 1-I. It is as if, in establishing an operational definition of temperature, we could do no better than to set up criteria which define which is colder or hotter, but were unable to develop quantitative relationships. Such considerations, even if they are valid, must await further progress in our understanding of the incredible variety of molecular interactions, linked functions, and correlated flows of matter, energy, and information, that one discovers when one examines living systems. In our view, it is premature at this time to attempt a physical definition of life on this route.

We shall take a short cut and consider only systems based on nucleic acids which self-replicate, mutate and compete. We do so without any claim to giving a definition with special merits, but only to delimit clearly our purpose here. We will concentrate on those phenomena which can be investigated at the molecular level, where physical principles can be unambiguously applied, and see how in a growing number of cases one begins to understand the physical basis for some of the distinctive features of living systems. The quotation at the beginning of the next chapter gives the viewpoint we shall endorse throughout this book.

TABLE 1-I

LIVING SYSTEM	DISTINCTIVE PROPERTIES
viroids	— self-replication by use of host enzymes, mutation — specificity to host — harmful but not destructive to host — individuality given by self-replication only, without compartmentalization
plasmids	— self-replication by use of bacterial enzymes, mutation — autonomous control of self-replication and distribution of copies — non-specific propagation within hosts — harmless or advantageous symbiont for host — individuality given by autonomy, functions without compartmentalization
viruses	— self-reproduction by use of whole host translation and replication machinery, mutation — individuality given also by compartmentalization — variety of shapes — specific recognition and ultimately destructive attack of host — latency by invasion of host genetic material
bacteria	— self-reproduction by individual translation and replication machinery, mutation — compartmentalization in cells — motility and response to external stimula — sexual mating

TABLE 1-I (CONTINUED)

protozoa

— self-reproduction by individual machinery mutation
— compartments inside cell with different functions
— motility and response to environmental stimula
— aggregation in colonies

higher organisms

— reproduction: sexual, germination, partenogenetic
— multicellular organization with cell differentiation
  for multiplicity of functions
— growth and morphogenesis
— motility in response to environmental stimula
— storage and elaboration of information about
  environment
— behavior
— social behavior
— self consciousness
— cultural evolution

CHAPTER 2

STRUCTURE AND ORGANIZATION OF LIVING SYSTEMS: AN OVERVIEW

> *There is still much lack of knowledge about this advanced level of organization of matter, we call "life", and its novel non-material consequences. However, at this stage it is simply "lack of knowledge", and not "discrepancy" with the present concepts of physics.*
>
> Manfred Eigen

The presently accepted view is that each of the variety of living systems present at this time on earth, as it appears now in its very special details or phenotypes, is the result of a process of evolution through selection in the competition for survival, "survival of the fittest" as proposed by Charles Darwin. Phenotypes are chosen to give an individual the best chances for survival only in relation to the particular environment which that individual's predecessors experienced. As the characteristics of the environment, which are of significance to living systems, may be strongly dependent on the activity of its predecessors and the existence of other living systems and of their activity, the process of evolution shows at least two time scales. When the environment changes slowly enough with respect to the characteristic time interval for reproduction of the living system under consideration, the phenotypes of this system evolve continuously toward the optimal. A classic example is the dominant color of butterflies in the vicinities of Manchester, which switched from white to dark grey when the city underwent the industrial revolution. The new color was obviously the fittest to escape being observed by predators when the environment darkened due to industrial pollution. When on the other hand the environment, or parts of it, changes too fast the steady evolutionary trend can be interrupted by collapses and instabilities and the intrinsic non-linear dynamics of the population of phenotypes, individuals, societies, etc., is fully revealed. It is just such a collapse which has been postulated to explain the demise of the dinosaurs and which could result from a nuclear exchange between super powers.

As one examines it, the concept of survival of the fittest may look like a tautology. We skip this discussion now, which sounds analogous to that on force and mass in Newton's second law of dynamics, but we shall examine in another chapter the model proposed by Eigen for molecular evolution, where different molecular species are seen to compete in such a way that a few dominate as the fittest and the fitness parameter is given a quantitative formulation in terms of chemical reaction affinities.

How then is a successful phenotype preserved and transmitted to the next generation? In recent times, molecular biology has given a remarkably full answer to this question, identifying at the molecular level the genes, which Mendel had hypothesized as the transmittable carriers of the information needed to maintain and/or steadily change phenotypes through subsequent generations. The elimination of genotypes promoting unsuccessful phenotypes is thought to occur as natural selection acts at the phenotypic level, selecting out unfit individuals, subspecies, species, etc.

The central dogma of molecular biology, in its original formulation, states that information flows exclusively from nucleic acids, the molecular carriers of genotypes, to proteins. Proteins in turn perform, or are responsible for the synthesis of other structures needed for the performance of virtually all chemical processes which take place in the living organism. Proteins comprise the enzymes which carry out chemical catalysis, including those involved in the energy transfer process of electron transport and photosynthesis. Other types of proteins are responsible for molecular recognition and specificity or for being structural elements. Among the reactions which proteins catalyse are the duplication of the genetic material so that it can be transmitted to the next generation, the reactions which result in its repair, and the reactions by which it is translated into proteins. Proteins are also involved in the control of those processes so that they occur at the appropriate time and rate. The informa-

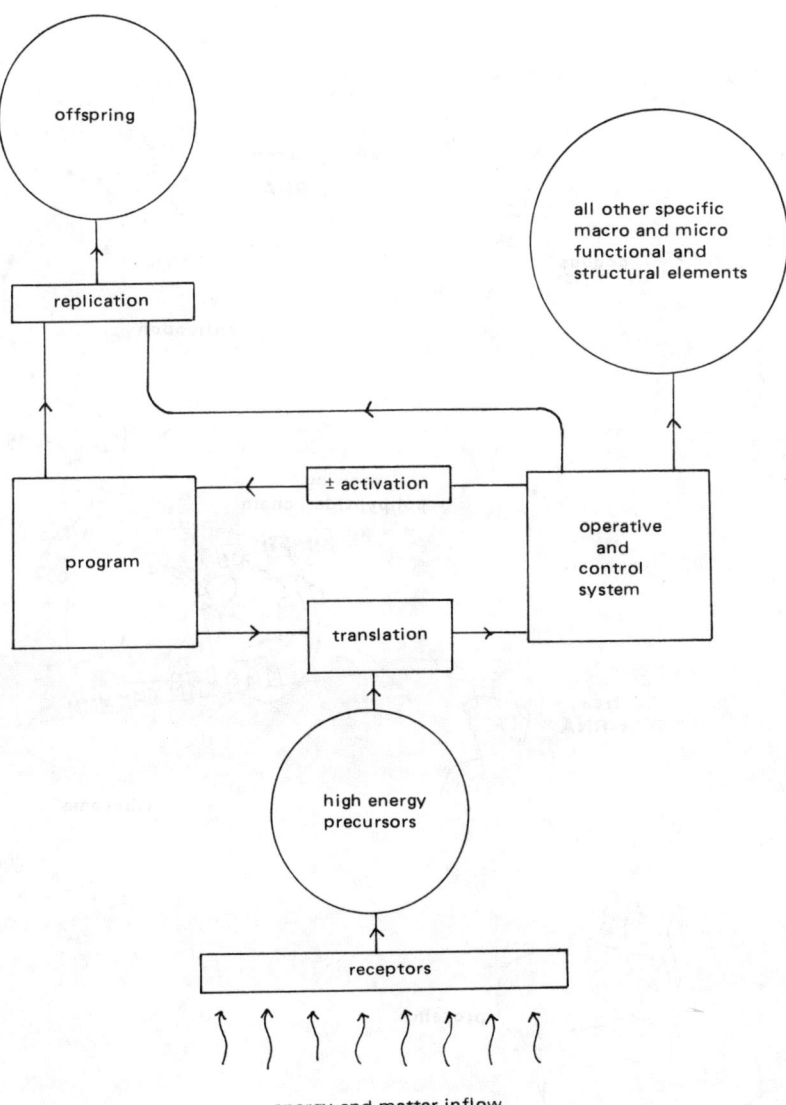

Fig. 2-1 A block diagram for "living systems". Those inside rectangular boxes are intended to have primary informative and functional significance, while products, intermediate, structural and other functional components are inside circles.

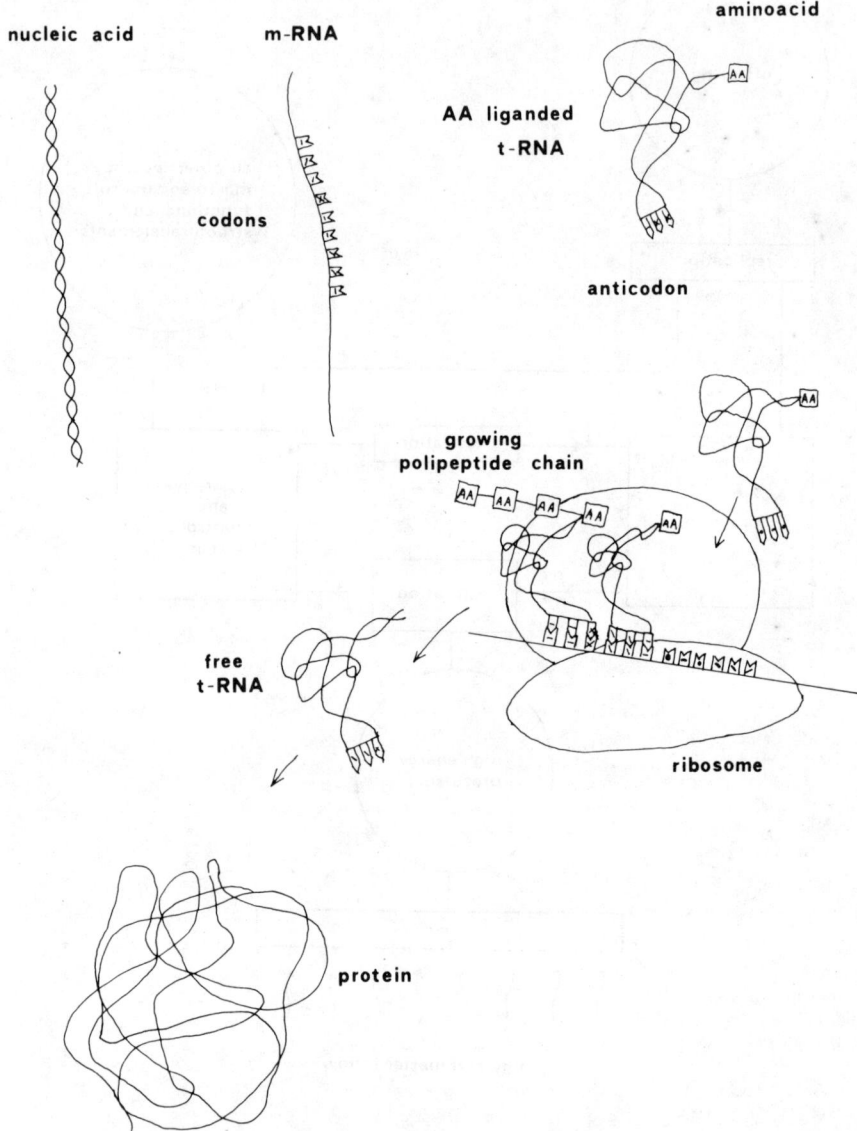

nucleic acid

m-RNA

aminoacid

**AA liganded
t-RNA**

codons

anticodon

**growing
polipeptide chain**

**free
t-RNA**

ribosome

protein

Fig. 2-2 Simplified schematic of the translation mechanism which
brings the information code in the linear RNA or DNA molecule into
the three-dimensional operative structure of a protein (see text).

tional and the operative molecular systems look circularly interlinked with a restriction in the direction of information flow, namely from nucleic acids to proteins. We give now an elementary treatment of how this is achieved, together with the properties of the relevant molecular structures. This is intended as a summary and a reminder of terminology, while an examination of the structures and functions of the molecules of life will be given in a subsequent chapter. An oversimplified block diagram is given in Fig. 2-1. Among actual living systems some will lack a few of the functions of the schematic and others will have additional ones. Those shown however are the essentials for living and the systems which lack some of them, as viruses, must "borrow" them from hosts which have them.

The "program" is coded in linear molecules, the single stranded ribonucleic acid, RNA, or the double helical deoxyribonucleic acid, DNA. The building blocks are monomers called nucleotides, of four different kinds. The specific sequence of the nucleotides codes the information to be read sequentially from a start signal to an end signal. The nucleotides are the letters of this alphabet and three of them in the sequence form a word. The set of triplets between a start and an end signal are a phrase and the set of all phrases along the whole linear molecule constitutes a book containing all the program for the living system. The smallest known RNA with self-replicating properties is only a few hundred nucleotides long. The largest DNA of a complex multicellular organism has hundreds of billions of nucleotides. The "translation" machinery uses high-energy precursors of nucleotides, and the assistance of enzymes, to translate a phrase of the book into RNA strands, the messenger RNA, which bring the information to the site of protein synthesis, the ribosomes. A protein is a specific sequence of aminoacids, linked one to the other by the strong peptide bond, chosen from the 20 aminoacids, occurring universally in living systems. A protein sequence is never branched, a fact that is consistent with its origin in the strictly linear, unbranched DNA or RNA. In the translation of the information a nucleotide triplet on DNA or RNA a word corresponds to one aminoacid. Single aminoacids, present in the aqueous environment, get bound to another nucleic acid of the translation machinery, the transfer RNA, t-RNA, which is of as many different kinds as the different aminoacids and, on a distinct molecular site, display a nucleotide triplet, the anticodon, which complements the codon specifying that aminoacid, as written on the m-RNA. The m-RNA binds to a ribosome, an assembly of proteins plus nucleic acids, which, being the site of protein synthesis, is the final part of the translation machinery. Here, the aminoacids carried by the various t-RNA's, are linked one to the other by processes assisted by specific enzymes, according to the m-RNA coded sequence. The linear chain of aminoacids spontaneously folds into the functional overall three-dimensional configuration, which, in that specific environment, is either the equilibrium conformation or a local minimum in free energy. One can immediately appreciate the relevance of the program + translation machinery, when one realizes that the biological function of each protein is mastered by its three-dimensional conformation and the specific conformation is mastered, through equilibrium or local equilibrium conditions, by the specific sequence with which the various aminoacids are arranged. Therefore for a protein of 100 aminoacids, its specific structure and function corresponds to choosing one, or at most a few, possibility out of 20^{100}. Fig. 2-2 gives a simplified schematic of the translation mechanism.

The "operative system" is given by the variety of proteins coded by different phrases in the book of the program. For instance the bacterium E. coli has about 3 000 different proteins of average molecular weight 40 000 accounting for 15% of its dry weight. According to the central dogma of molecular biology, the flow of information is unidirectional, from nucleic acids to proteins. On the other hand all operations pertain to the proteins. They operate the activation or repression of the program, so that the translation of each protein is timed in such a way to get it when and in the amounts needed. Proteins promote as activators and assist as catalysts the self-replication of the program: DNA and RNA instruct, the synthesis of complementary strands from energy rich monomers to produce a replica of the original but have no means to operate without specific enzymes. The replica is given as the program to the new generation. Finally proteins participate in the bioenergetics, in the synthesis of the molecular building blocks of the living system, in the exchange of molecular messages within and outside the living systems, etc. Finally, one should stress their properties as molecular regulators, not only in the processes involving nucleic acids, but in all biochemical pathways of a living system when control is needed. We shall not enter into this now and rather end this short review here.

Before starting to study in detail the properties of living systems, it may be useful to see an overview of their diversity in structure, complexity, and organization. We shall do this with the aid of a somewhat arbitrary choice of pictures, which will also be an occasion for starting to extend our understanding of the basic mechanism just described.

We emphasize again that the various living systems, although presented for convenience in an order of increasing complexity, higher organization and richer behavior, can by no means be considered to differ in their evolutionary sophistication. Each is as fit, the fittest to survive in its environmental niche.

Viroids are the smallest molecular aggregates that replicate, keeping intact their identity through subsequent generations, except for mutation and population variability. In Fig. 2-3 an electron micrograph of PSTV viroids is shown. Each individual is a strand of RNA, incredibly short: only 369 nucleotides for PSTV. The RNA strand can be linked in a circle or be simply linear. The native conformation shows short regions of base pairing and double helix formation alternating with even shorter unpaired regions of single stranded RNA. The different conformational states depicted in Fig. 2-3 have actually been found in electron micrographs of partially denatured viroids.

Until now, viroids have been observed as pathogens only in plants, but they may also be pathogens for animals and perhaps man. They are of different species as demonstrated by the fact that viroids from one plant preserve their own characteristic nucleotide sequence when propagated in a different plant.

How do they manage to replicate, which is apparently their only living activity? This is still not known in detail. It is rather known how they do not do it. The 359 nucleotides of PSTV are too few to code for a specific protein, which would help in replication. If one would allow for up to three rounds of translation, which, given the known sequence, are potentially possible for the circular arrangement of an uneven number of bases, one sees seven possible initiator triplets of bases and six possible terminator triplets. Then one may expect any translation product, from a tetrapeptide to a protein, resulting from more than two rounds of translation. However the sequence lacks binding sites for ribosomes at the possible initiator site and this, along with other reasons, seems to preclude translation into peptides or proteins specific for the viroid. Moreover in infected plants, no proteins are found which are not present in uninfected plants. Therefore, it appears that the viroid RNA does not act as a messenger RNA coding for a specific protein. Viroids may simply replicate with the aid of enzymes already present in the host plant, without interfering with the translation machinery of the host, as viruses do. Indeed, enzymes, that can replicate RNA using RNA templates, have been found in some plants. There is a well documented example of an RNA system which replicates in vitro in the presence of only nucleotide triphosphates and a specific enzyme. An RNA strand of some 220 nucleotides, which is obtained as a free section of the 4 500 nucleotide RNA of the $Q\beta$ virus, replicates in vitro with only the aid of a specific enzyme, the $Q\beta$-replicase, as long as the four monomeric precursors of RNA, in the activated triphosphate form, are present. We shall come back to the $Q\beta$ system in another section of this book. We mention it here only to show that such a "simple" replication mechanism

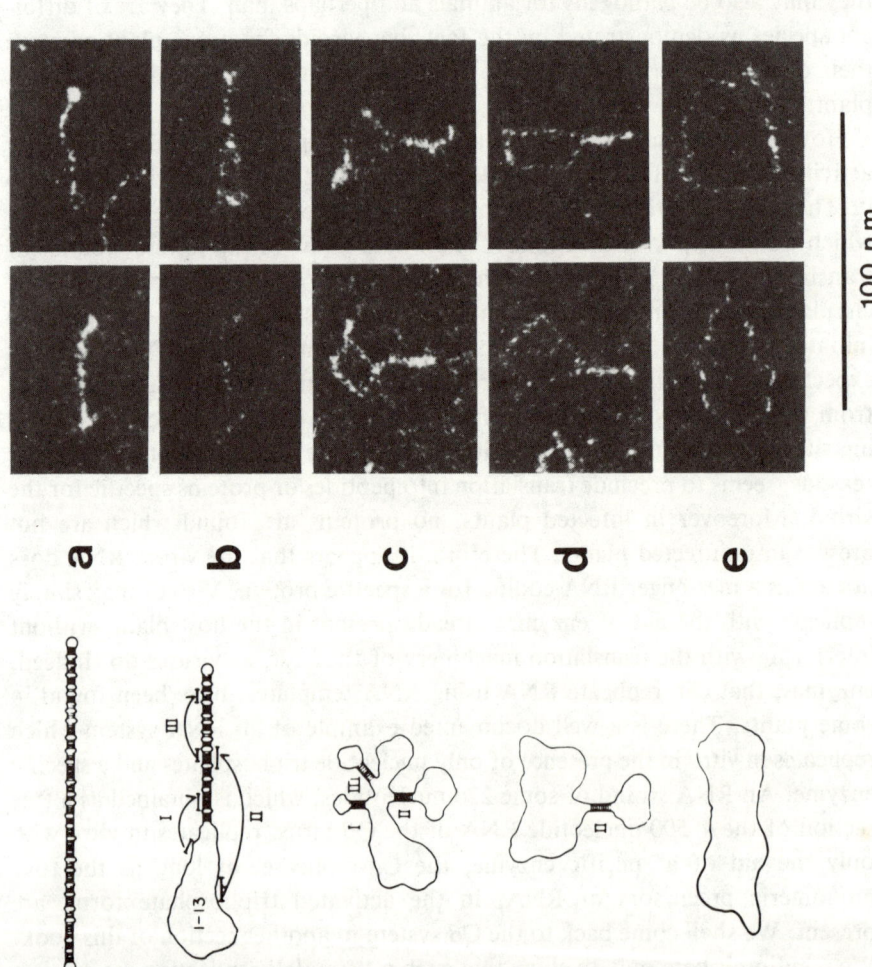

Fig. 2-3 Electron micrographs of various conformational states of PSTV viroids; the black bar, equivalent to 10³ Å, gives the scale; the schematics guide to the micrographs. The RNA shows sections of base pairing in double stranded helical conformation and sections of single stranded segments, which may constitute larger and larger parts of the molecule (top to bottom) as in various conformation states of the molecule.
(Reproduced with permission from: G. Klotz and H. L. Sanger, Eur. Journ. Cell Biol. 25, 5 (1981)).

100 n m

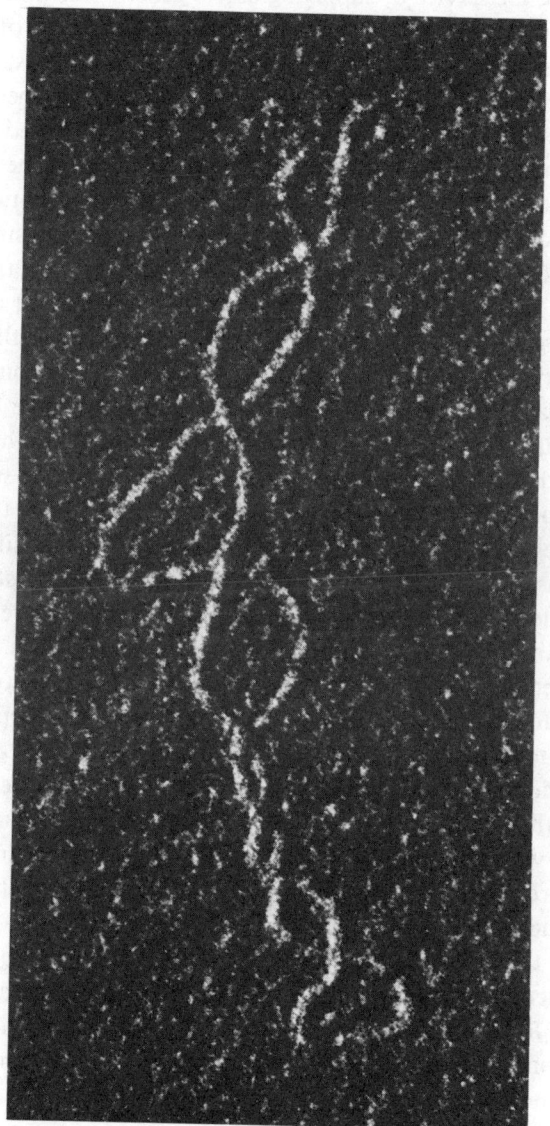

Fig. 2-4 An electron micrograph of the plasmid pBR322 from E. coli, consisting of 4 362 nucleotide pairs; notice the superhelical arrangement of the molecule (see also Fig. 2-14); magnification × 250.000.
(Reproduced with permission from: presentation to 8th European Congr. Electron Microscopy 1984, courtesy of G. Klotz).

is quite possible and could be that used by viroids to "live".

Plasmids are endosymbionts, which live stably in the host bacterial cell, sometime performing the service of supplying useful genetic functions such as resistance to toxic agents. They are strands of double helical DNA of characteristic length of tens of thousands of bases, Fig. 2-4. They can exist within a cell in a number of copies of each of various kinds. The copy number of each kind ranges from just one up to some one hundred and is conserved through subsequent generations. Replication occurs randomly throughout the host cell cycle. The mechanism which controls the conservation of the copy number is sometimes unable to distinguish between two closely related plasmids and therefore conserves the sum of their copy numbers. Plasmids carry the information for bacterial mating, see Fig. 14 for bacterial sex in action, and they can be exchanged together with the chromosomal genes. However they show their full individuality as living systems as i) they are exchanged in conjugation between different bacterial species, which are unable to exchange chromosomal genetic material, ii) they can transfer from one species to another which is in competition with the first for survival, enabling the recipient to survive at the expense of the donor. Together with the autonomy in controlling their own copy number, these features characterize the plasmid's way of life as independent organisms, able to survive whatever happens to their host.

We have seen the least demanding forms of living systems: just a replicating linear molecule, with enough information encoded in its sequence to allow control of its self-replication. There is very little recognition, if any, of the host, on whose enzymes and nucleotide precursors they depend for replication: viroids can propagate in plants different from those in which they are originally found, plasmids seem to survive independently of the specific host bacterium in which they happen to live. Indeed both viroids and plasmids tend to be in symbiosis with the host. They have no interest in recognizing it, attacking it or destroying it, as probably they would have difficulty, these naked strands of nucleic acid, to preserve their integrity outside the host. It should be said that these forms need not be primordial in any respect: they may well be an outcome of comparatively late evolution. For instance they could be strands of RNA and DNA, which had come loose from the genetic material of more complex living systems and found suitable enzymes and a favourable environment for self replication in their hosts and then acquired the capability of surviving also in other hosts.

The way of life of viruses is quite different. First of all they are predators, at a very fundamental level: they appropriate in full the biochemical machinery, which their prey uses for synthesis of its proteins and replication of its genetic material. That machinery, once switched under the control of the viral genetic material, will work no more in pace with the metabolism of the prey, but much faster until all the resources of the prey, in terms of nucleotides and peptides, precursors of nucleic acids and proteins, are used up and a large number of copies of the original virus have been formed. The prey cell undergoes lysis, but there are few problems for the free viruses. In the extreme, when they find themselves in a biologically inert medium, they can stay unharmed in a wide range of solution conditions or they may even crystallize and therefore enter unharmed dry and solid phases, ready to resurrect again when in touch with living cells. The trick is played by having the viral genetic material packed in and protected by a coat of protein. The coat of proteins is specific for the virus strain and is coded for by the viral genetic material. Fig. 2-5a shows a schematic of the simplest virus: it is an RNA virus, a strand of RNA some 3 500 bases long, coiled and encapsulated in a spherical protein coat of dimensions 500 Å . The protein can fuse into the membrane of the cell to be attacked and then the viral RNA can enter the cytoplasm and start the reproductive cycle of the virus.

In some cases the virus can show latency. Rather than get started right away, the viral genetic material can get itself incorporated into the genetic material of the cell, be replicated with it through subsequent generations and then revert to activity in response to external stimuli such as light, temperature, etc. This is an example of control of gene expression. Fig. 2-5b gives a schematic of both types of viral life cycles.

Viruses can be much more complex structures. Apart from the geometrical regularities in their capsids, which allow crystallization or in general regularities in aggregation, they show substructures of functional significance. Fig. 2-6a shows the T4 phage. The virus is provided with a recognition site, prongs and coils to recognize the prey membrane, punch it and inject the DNA in the cell, as in the schematic of Fig. 2-6b. One may wonder how all these different substructures are assembled after they have been synthesized at the expense of the prey cell. An elegant experiment showed that it is somewhat of a spontaneous process in the proper ambient conditions: virus strains, each defective in promoting the synthesis of different substructures, produce material which, when incubated together, will reassemble by mutual specific molecular interactions to give the intact, fully infective form.

The capsid of a virus is only for protection, none of the processes of the life cycle of the virus occur in its interior. Cells give a different solution to the problem of individuality and survival: there is metabolic activity in the inside of the lipid membrane which constitutes their boundary. Fig. 2-7 gives a

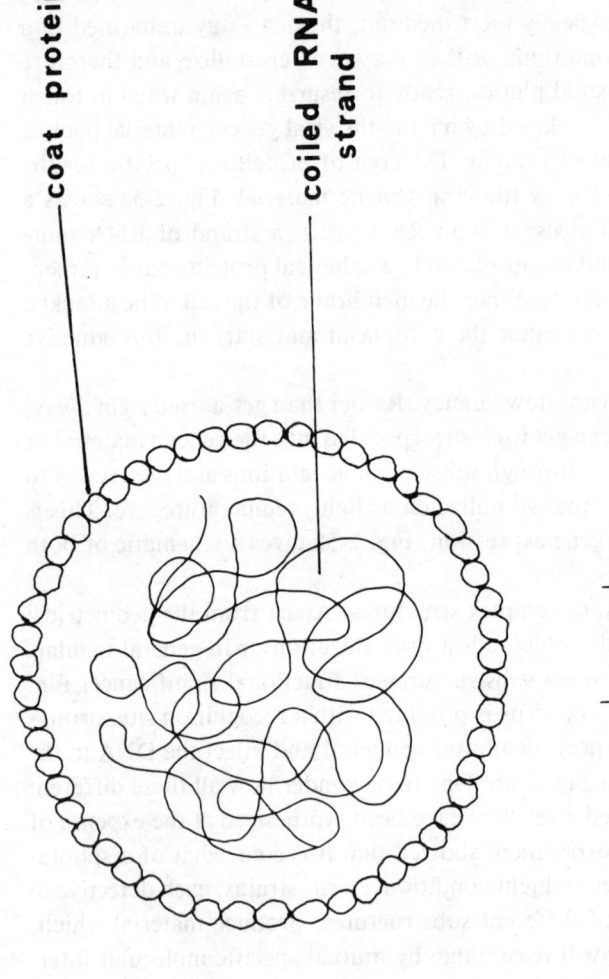

Fig. 2-5   Schematic of a simple RNA virus (a) and of viral life cycle (b). After the virus particle fuses into the bacterial cell, two paths can be taken: (left) replication of a number of copies of the virus genetic material and coat protein, which reassemble and burst out of dead bacterium or (right) insertion of virus genetic material into the bacterial gene and replication of the composite gene through subsequent bacterial replications. External stimuli as light, higher temperatures, etc. may interrupt such a latency and revert to activity.

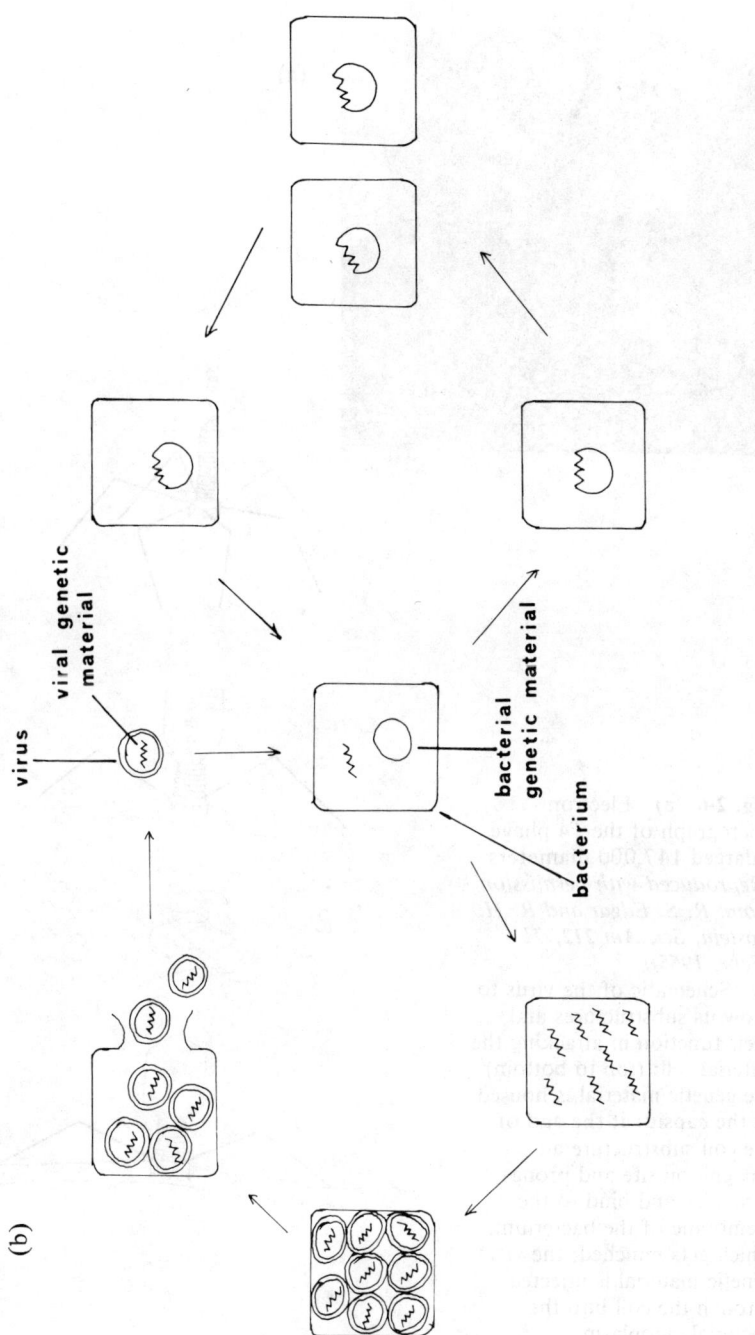

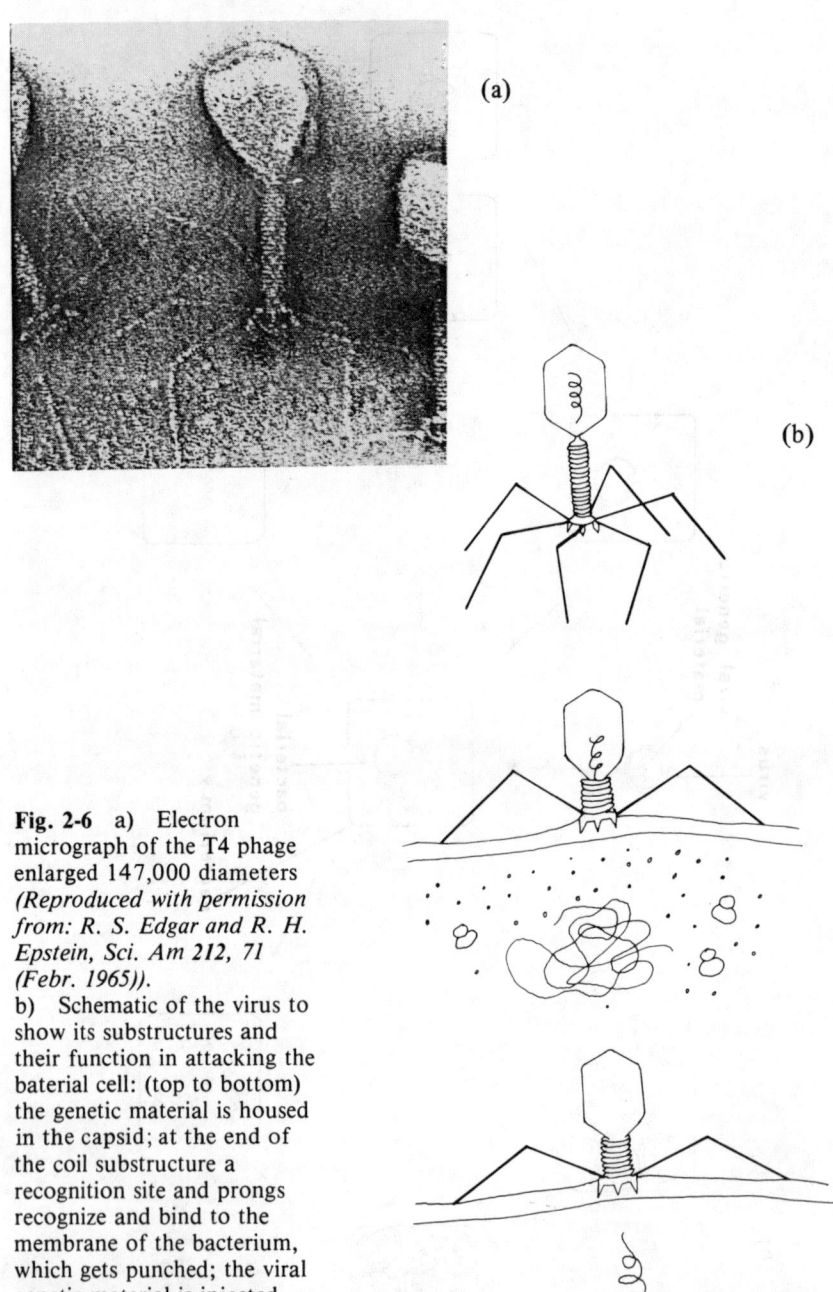

Fig. 2-6   a)   Electron micrograph of the T4 phage enlarged 147,000 diameters *(Reproduced with permission from: R. S. Edgar and R. H. Epstein, Sci. Am 212, 71 (Febr. 1965)).*
b)   Schematic of the virus to show its substructures and their function in attacking the baterial cell: (top to bottom) the genetic material is housed in the capsid; at the end of the coil substructure a recognition site and prongs recognize and bind to the membrane of the bacterium, which gets punched; the viral genetic material is injected through the coil into the bacterial cytoplasm.

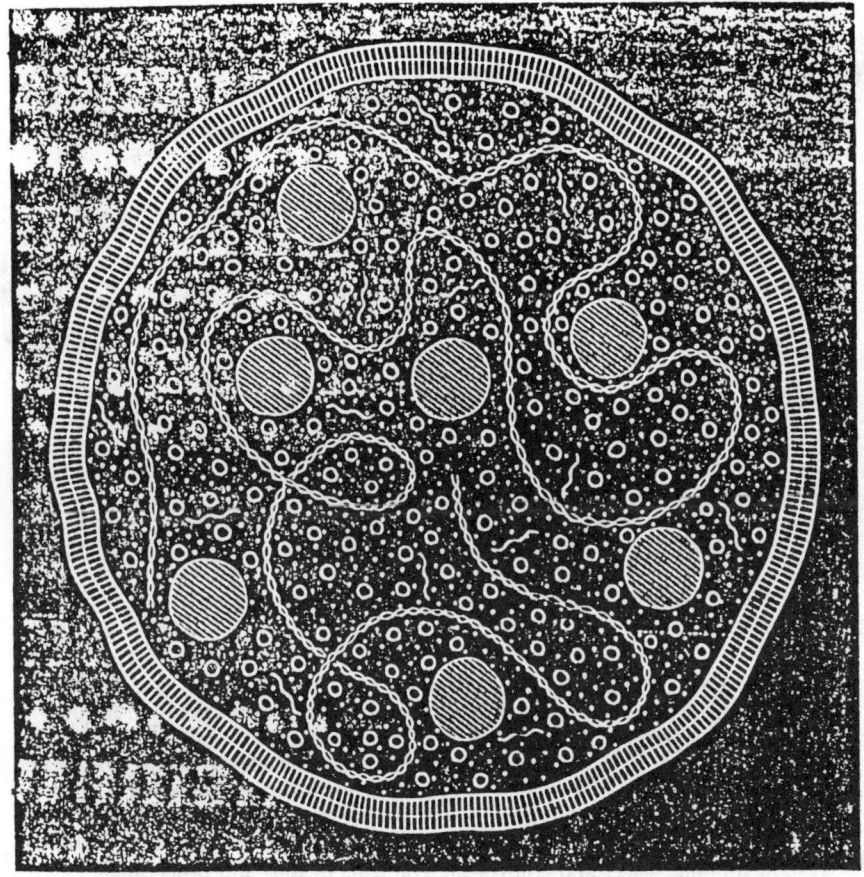

Fig. 2-7 Schematic representation of the smallest cell, a cell of Pleuro Pneumonia like Organism, PPLO. The size is approximately 10^3 Å, with a weight of 5×10^{-16} g. For comparison a protozoon would show a size of 10^6 Å and a weight of 5×10^{-7} g. Key: (▥) lipoprotein membrane, (●) ribosome, (○) soluble protein, (∿) soluble RNA, (⋙) DNA, (•) metabolite.
(Reproduced with permission from: H. J. Morowitz and M. K. Tourtellotte, Sci. Am. 206, 116 (March 1962)).

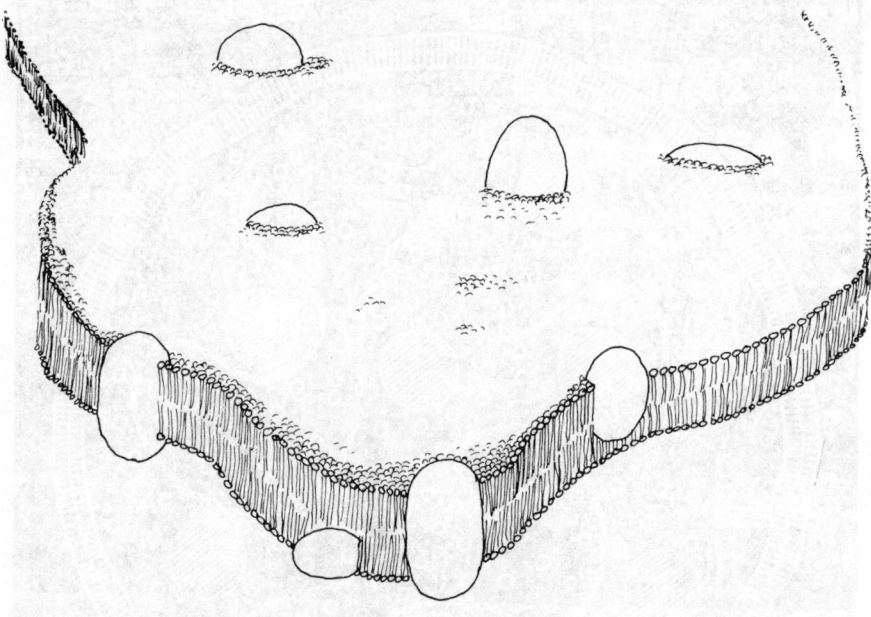

Fig. 2-8   Schematic of a cell membrane: a bilayer of lipids encloses the cell; their polar heads are directed towards the aqueous medium inside and outside the cell; the bilayer is of typical thickness   ~70 Å; it is impermeable to water, ions and other small molecules; proteins fuse in the lipid bilayer, with freedom of lateral diffusion, and either cross it or enter it in part, facing either the outside or the inside, or stick either to the inside or the outside. The proteins perform all the functions of the membrane: select and control the gradients of ions and other small molecules, provide sites for cell-cell recognition, for receptors, etc.

schematic of the smallest and simplest living cell, the so called PPLO, Pleuro Pneumonia Like Organism. The genetic material, which is DNA for any cell, is expressed in full autonomy: the cell contains the translation machinery for DNA replication and ultimately is able to duplicate as a whole. The need to create an interior is not limited to protection only. There is the need to keep all the molecular equipment in an aqueous environment suitable for metabolism, where enzymes, mRNA, tRNA, ribosomes, DNA etc. can easily meet and also find a proper concentration of precursors, either fabricated inside the cell from even less specialized molecules or directly found in the external medium. Ordered structures of lipids constitute the membrane separating interior from exterior. Specific proteins and small peptides can dissolve in this two dimensional fluid where they can move laterally and are responsible for the selective permeability to and the active transport of the molecules needed for metabolism. Fig. 2-8 shows a schematic of a biological membrane.

What is there and what happens in a cell? In a rapidly dividing cell of the bacterium Escherichia coli, which weighs approximately 10^{12} daltons, one finds 70% by the weight of water, 15% by the weight of proteins representing 3 000 different kinds, of average molecular weight 40 000 daltons; DNA, present of course only in one variety of molecular weight, 2.5×10^9 daltons, accounts for 6% of the weight together with the other nucleic acids involved in the translation machinery, mRNA, tRNA, rRNA; precursors and molecules of carbohydrates, lipids, nucleotides, and aminoacids account globally for about 6% of the weight, each distributed between a few hundred different kinds; finally 1% of the weight is given by about 20 different kinds of inorganic ions, which include Na^+, K^+, Mg^{a+2}, Ce^{a+2}, Cl^-, PO_4^{-4} and Fe^{+2}. All the components diffuse in the cytoplasm, in particular the DNA, with the exception of the enzymes of the respirating chain which stick on the inside of the inner cell membrane, which controls permeability. Fig. 2-9 shows a highly simplified schematic of the metabolic pathways at work in living E. coli, which are almost entirely known at present. Two properties are immediately evident: the metabolic pathways need to be driven by chemical energy, aerobic metabolism in this case, and they are all interlinked.

Let us notice that such a linkage calls for some kind of regulation, in order for the biochemical machinery to work in phase. None of the pathways can be permitted to run away, but all of them must be in delicate tune with each other, producing just enough of the peptide, cleaving just enough of those high energy phosphate bonds, translating DNA into that protein just so many times an hour etc. Biochemical feedback is achieved at the molecular level by proteins designed for that purpose, of which there need be only a few in a metabolic

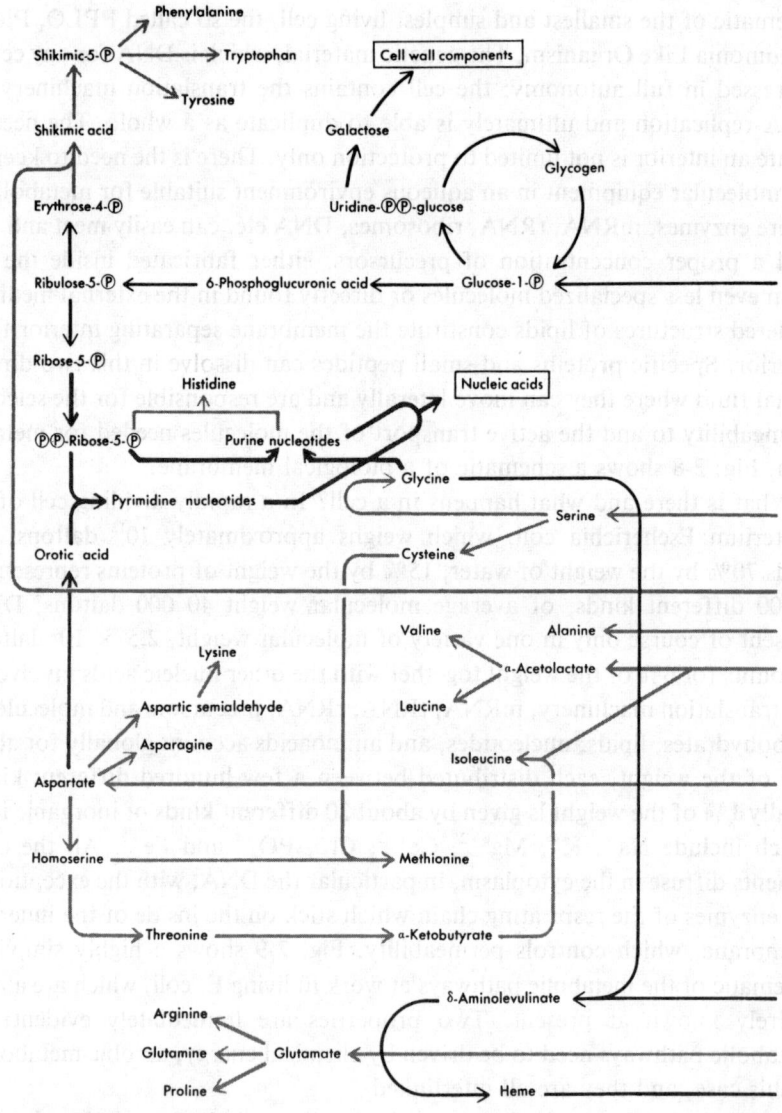

Fig. 2-9   Schematic view of some of the main metabolic pathways in
E. coli.
*(Reproduced with permission from: J. D. Watson, "Molecular biology
of the gene", 3rd ed., Copyright © 1976 The Benjamin/Cummings
Publishing Co. Inc.).*

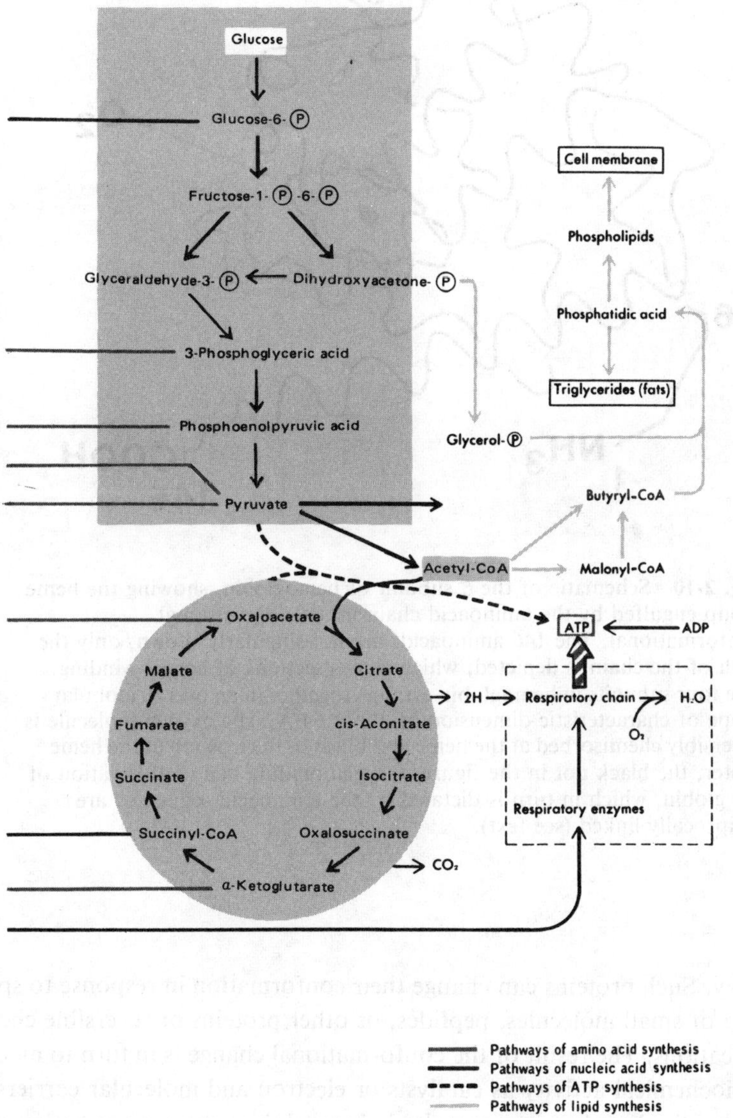

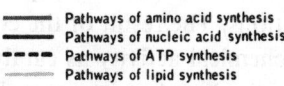

Glucose
Glucose-6-Ⓟ
Fructose-1-Ⓟ-6-Ⓟ
Glyceraldehyde-3-Ⓟ
Dihydroxyacetone-Ⓟ
3-Phosphoglyceric acid
Phosphoenolpyruvic acid
Pyruvate
Acetyl-CoA
Malonyl-CoA
Butyryl-CoA
Glycerol-Ⓟ
Triglycerides (fats)
Phosphatidic acid
Phospholipids
Cell membrane
ATP
ADP
2H
Respiratory chain
H_2O
O_2
Respiratory enzymes
Oxaloacetate
Citrate
Malate
cis-Aconitate
Fumarate
Succinate
Isocitrate
Succinyl-CoA
Oxalosuccinate
α-Ketoglutarate
CO_2
Pathways of amino acid synthesis
Pathways of nucleic acid synthesis
Pathways of ATP synthesis
Pathways of lipid synthesis

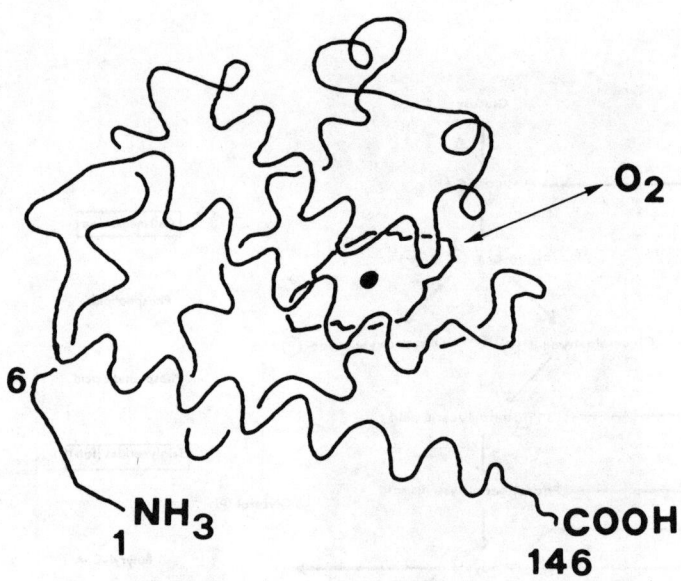

Fig. 2-10 Schematic of the β subunit of hemoglobin, showing the heme group engulfed by the aminoacid chain in its tridimensional conformational. The 146 aminoacids are not singularly shown, only the path of the chain is depicted, which shows sections of helical winding. The four subunits in hemoglobin arrange together in an overall globular shape of characteristic dimension of about 64 Å. The oxygen molecule is reversibly chemisorbed at the heme and binds to the iron ion at the heme center, the black dot in the figure. Oxygen binding and conformation of the globin, which in turn is dictated by the aminoacid sequence, are reciprocally linked (see text).

pathway. Such proteins can change their conformation in response to specific binding of small molecules, peptides, or other proteins or reversible chemical modifications. The result of the conformational change is in turn to modulate their biochemical activity as catalysts or electron and molecular carriers etc.

Among the best known examples is hemoglobin, the oxygen carrier in the blood of a few invertebrates and all vertebrates. Four hemes, one in each of four subunits, with an iron atom at the center of the heme, can reversibly chemisorb molecular oxygen. This process promotes a conformational change in the structure of the subunit, which alters binding at the contact between subunits and, as more and more subunits are oxygenated, ultimately results in a change of the conformation of the whole molecule. This in turn forces any subunit still unoxygenated into a conformation in which it is easier for its heme to bind oxygen. By this trick the affinity for oxygen binding of the last heme can be larger by a factor of a few hundred than that of the first heme. Moreover, the overall oxygen affinity is controlled by the interaction of small molecules with other parts of the protein. The whole of these responses to solution conditions allows for the process of respiration at a higher efficiency and the possibility of adaptation to different environmental conditions. Fig. 2-10 shows a schematic view of the β subunit of the hemoglobin molecule. It is worth noticing at once that quite a bit of this bulk structure is crucial to the biological function: the substitution of just one aminoacid, glutamic acid with valine, in a specific spot, position 6, still quite far apart from the heme-iron complex, alters drastically the solution properties of the protein, giving rise to a disease, sickle-cell anemia. This is a genetic disease, as the wrong aminoacid, in the sequence of some 140 in each subunit, is present as a consequence of an error in the DNA sequence, the gene, which directed the synthesis of the subunit. So one can fully appreciate how linear information in the genetic material, translated into the tridimensional conformation of proteins, produces specific biological functions.

Before we move to the next scale of complexity, let us examine more fully the way of life of bacteria. They can move around, in response to gradients of chemicals, chemotaxis, of light, phototaxis etc. How they get the message, and how they translate it into a reaction of motion is still largely unknown. What is better understood is how they move, once they have decided to do so. They whirl a helical filament at a hundred turns per second, while their body counter-rotates at some ten turns per second. The filament is a rigid protein structure attached to the cell wall by means of a connector, which is a wonderful molecular miniature on a nanometer scale of a rotary joint made of hook, rod, shaft and rings. Fig. 2-11 shows micrographs and a schematic. The rotor could be activated in rotation in respect to the stator by some kind of translocation of ions on the rotor, but details are still unknown.

Bacteria may enjoy sex, as Fig. 2-12 shows. The sex factor is carried by plasmids, which we have already encountered as symbionts. The sexual activity is of utility to both, as it allows transmission of the antibacteria resistant factors carried by the plasmids and it allows growth in the total population

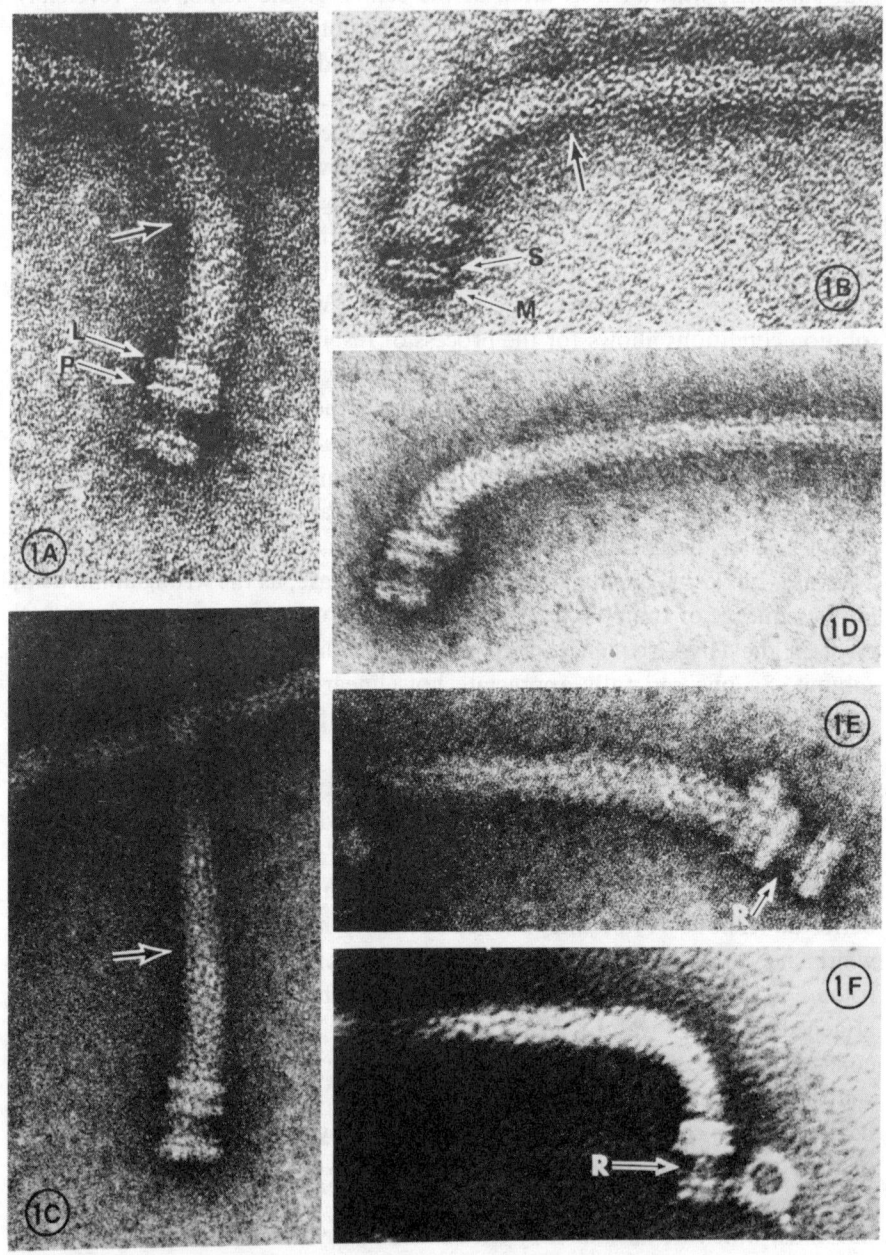

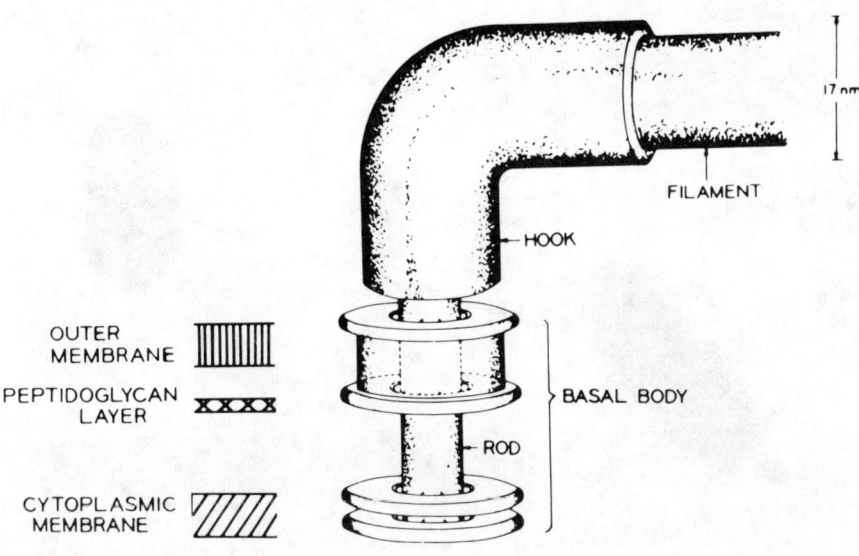

OUTER
MEMBRANE

PEPTIDOGLYCAN
LAYER

CYTOPLASMIC
MEMBRANE

FILAMENT

HOOK

BASAL BODY

ROD

17 nm

Fig. 2-11 Basal end and intact flagellum from E. Coli seen in various micrographs (magnifications of the order of 315.000 diameters; bars in each part of the figure represent 30 nm); arrow marks the junction between hook and filament; R marks the rod connecting top and bottom rings; L, P, M and S mark the four rings in the basal body. The various structural elements seen in the micrographs are shown schematically in the drawing.
(Reproduced with permission from J. Adler, J. Bacteriology 105, 384 -395, (American Society for Microbiology, 1971)).

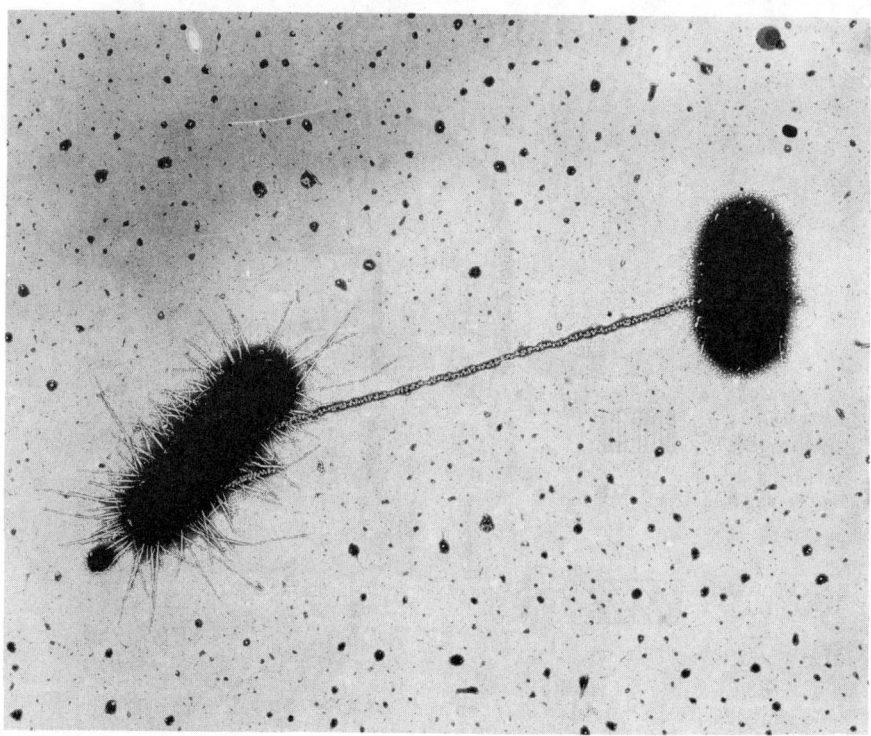

Fig. 2-12   Sexual mating in E. coli bacteria: the "male" (left) transfers
the genetic material to the "female" (right) using the long pilus, which
joins temporarily the two.
*(Reproduced with permission from: R.Y. Stanier, M. Doudoroff and
E. A. Adelberg "The microbial world", 3rd ed., Prentice Hall (1970);
electron micrograph prepared by J. Carnahan and C. Brinton).*

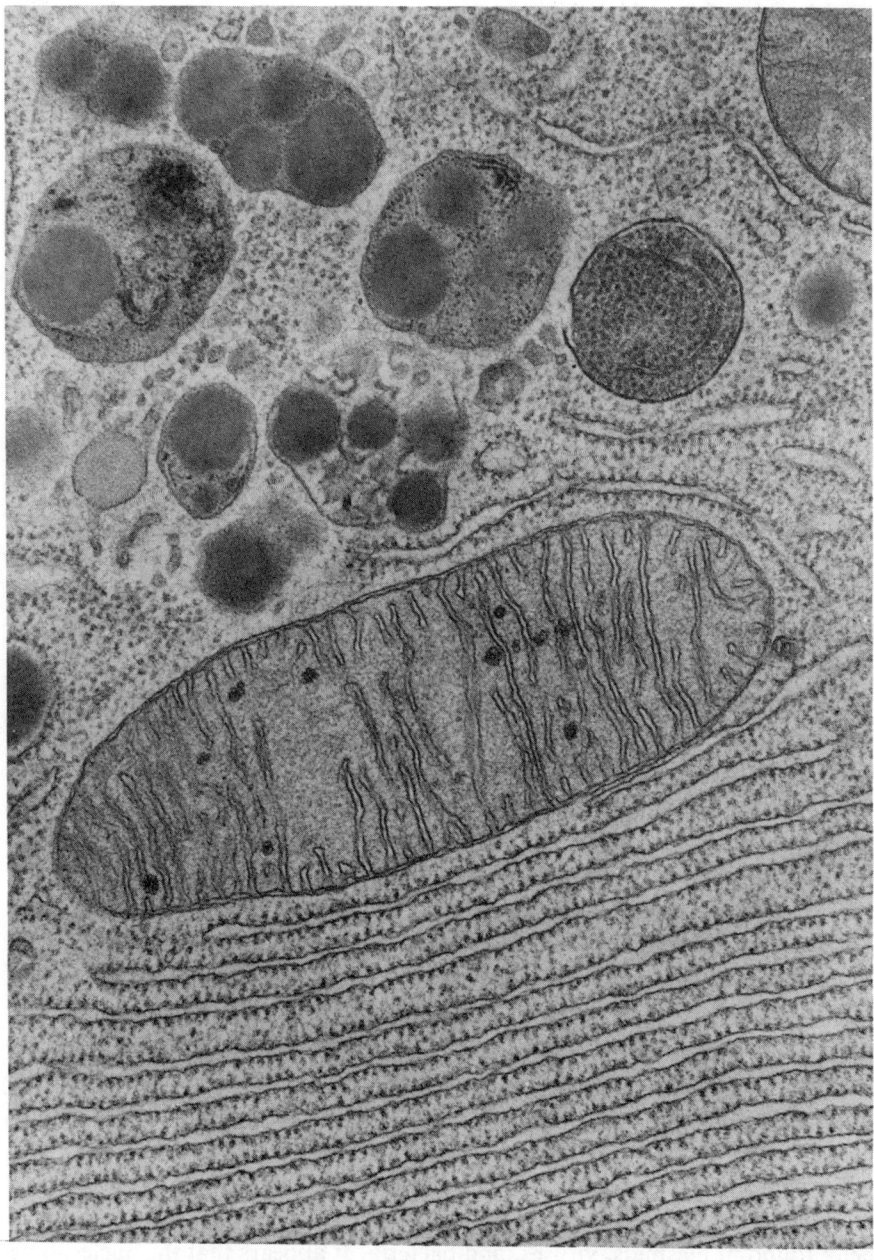

Fig 2-13 Mitochondrion from a bat pancreas cell, magnification × 48,360. See also the schematic Fig. 5-6 for details of the internal substructures. (*Courtesy of K. R. Porter*)

for the plasmids, which by the process colonize bacteria formerly lacking them.

Is the mitochondrion, the organellum where respiration occurs in eukariotic cells, a progenitor oxygen respirating bacterium, which troubled primordial anaerobic cells learned to live on and with, accepting it as a symbiont? It will be difficult to establish such a notion beyond doubts. Still it is interesting to note that mitochondria code with their own DNA for a few of their proteins, replicate in syncrony but independently upon cell division and their own genetic code is slightly different from the universal code of contemporary cells and bacteria, an unique exception. Fig. 2-13 shows a micrograph of a mitochondrion.

In plasmids, viruses, bacteria and mitochondria the DNA may be organized in a superhelical arrangement: the double helix winds up to form a helix of higher order, Fig. 2-14. The properties of these structures and the relations to biological function are studied with the mathematical methods of topology and of differential geometry. The topological characteristics of a DNA molecule in a living system are not immutable, but can be changed by the intervention of special enzymes, the topoisomerases.

Eukariotic cells, that is cells in which the genetic material, DNA, is housed in a nucleus delimited by a membrane, are characteristic of protozoa, algae, fungi and all plants and animals. They are quite complex structures, as shown in the schematic of Fig. 2-15, which does not depict any specific cell, but lists what is understood presently as the typical composition and structure of all cells. To the rich variety of organella with various functions, one would add more for specific organisms, as cilia for motion in protozoa, molecular structure to develop pseudopodia for motion in amoebae and many other cells, etc.

Recently it has been found that cells display also a microtrabecular structure, which keeps other organella, microtubules, polyribosomes, endoplasmatic reticulum, etc. in a recticular arrangement. Fig. 2-16 shows a schematic. The interior of the cell is no longer a simple solution in which organella swim more or less freely, but rather the cytoplasm is organized by the microtrabecular reticulum, to which enzymes can be attached. Small molecules in the aqueous phase then find shorter average distances between successive encounters with enzymes and the biochemical machinery may work faster.

The genetic material in the nucleus of eukariotic cells is organized in a much more complex architecture than the simple strands of double helix and superhelix found in plasmids, viruses, and bacteria. It appears that the nucleosome is the fundamental building block. It is an association of DNA with specific proteins, called the histones, where the DNA, organized as a superhelix, regu-

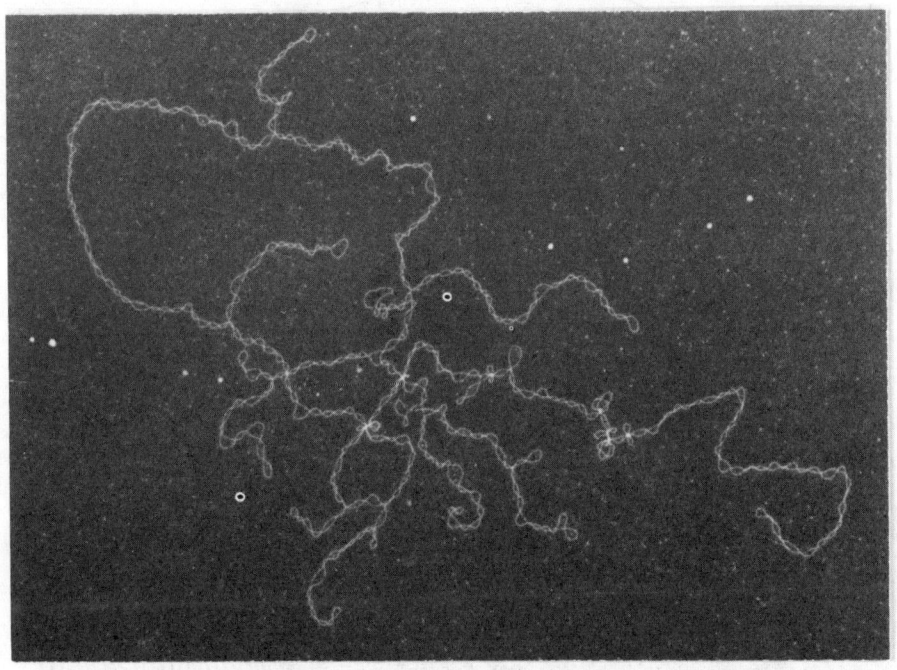

Fig. 2-14 Electron micrograph of superhelical DNA of a large plasmid, MW 10^8, from Halobacterium Halobium; magnification × 26.400. *(Reproduced with permission from: G. Weidinger, G. Klotz and W. Goebel, Plasmid 2, 377 (1979) Academic Press Inc.)*

larly coils on a cylindrical arrangement of four of the histone proteins, the fifth somewhere closing the structure. How these substructures give rise to the much more complex chromosomes and how sections of DNA are selected at the appropriate times to be unwrapped from the more complex structure and exposed for transcription, Fig. 2-17, is still a matter of active research.

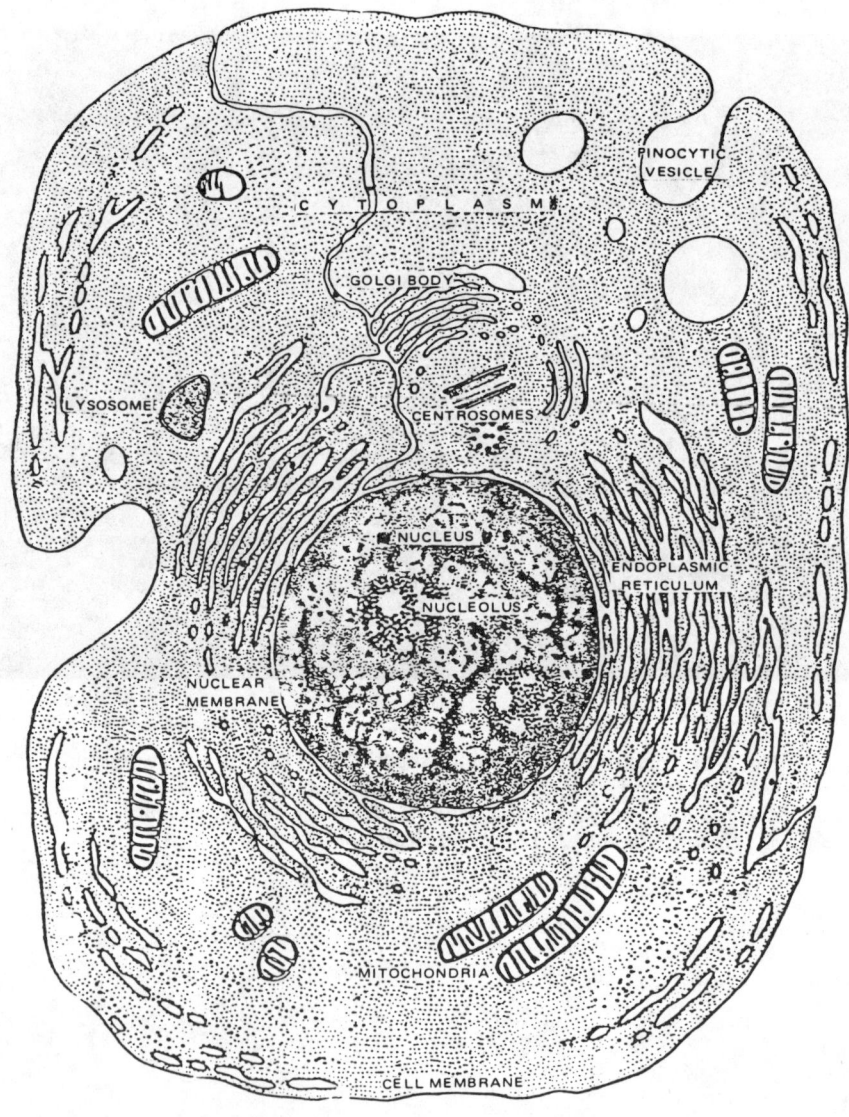

Fig. 2-15   Schematic of the typical eucariotic cell (see text). The dots that line the endoplasmic reticulum are ribosomes.
(Reproduced with permission from: J. Brachet, Sci. Am. 205, 51 (Sept. 1961)).

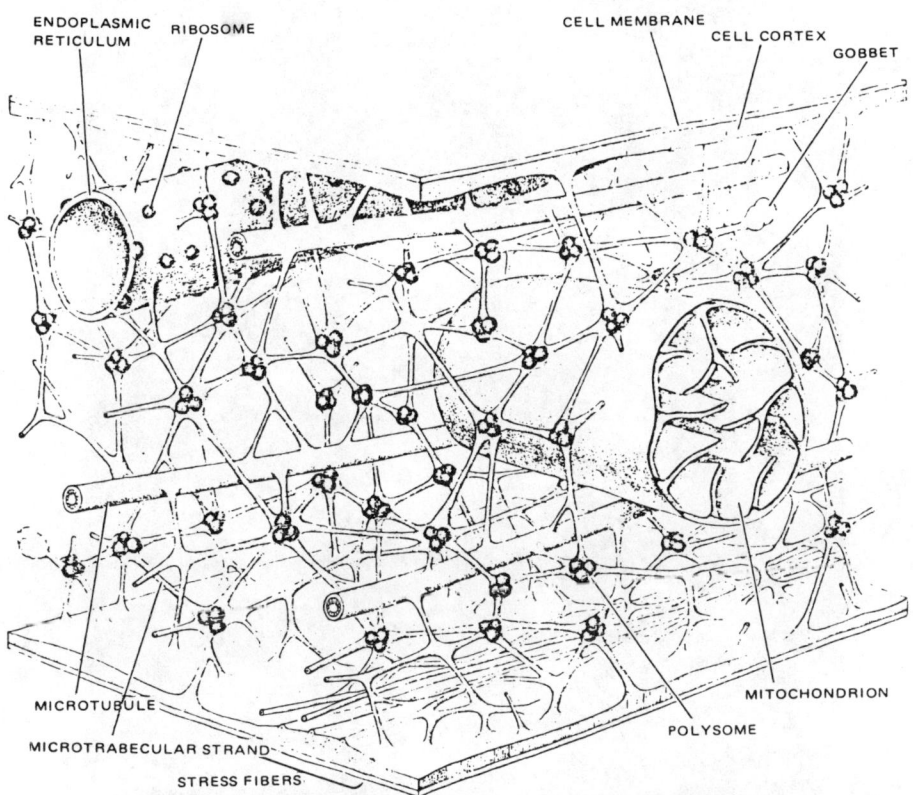

ENDOPLASMIC RETICULUM   RIBOSOME

CELL MEMBRANE   CELL CORTEX   GOBBET

MICROTUBULE

MICROTRABECULAR STRAND

STRESS FIBERS

MITOCHONDRION

POLYSOME

Fig. 2-16 Model of the microtrabecular lattice, some x 204,000 its actual size, was derived from hundreds of images of cultured cells viewed in the high-voltage electron microscope. The model illustrates how the microtrabecular filaments are related to the other components of the cell cytoplasm: the substance of the cell outside the cell nucleus. In the model the microtrabeculae suspend the elongated structures of the endoplasmic reticulum (the system of interconnected channels within the cell where some of the proteins manufactured by the cell are sequestered), the mitochondria (the organelles that manufacture ATP, the universal fuel of the cell), the microtubules (the complex fibers that serve many functions of cell structure) and the microfilaments buried in the cell cortex (the layer of material just under the outer membrane of the cell). At junctions of the microtrabecular lattice are polysomes: organized clusters of ribosomes.
(Reproduced with permission from: K. R. Porter and J. B. Tucker, Sci. Am. 244, 41 (March 1981)).

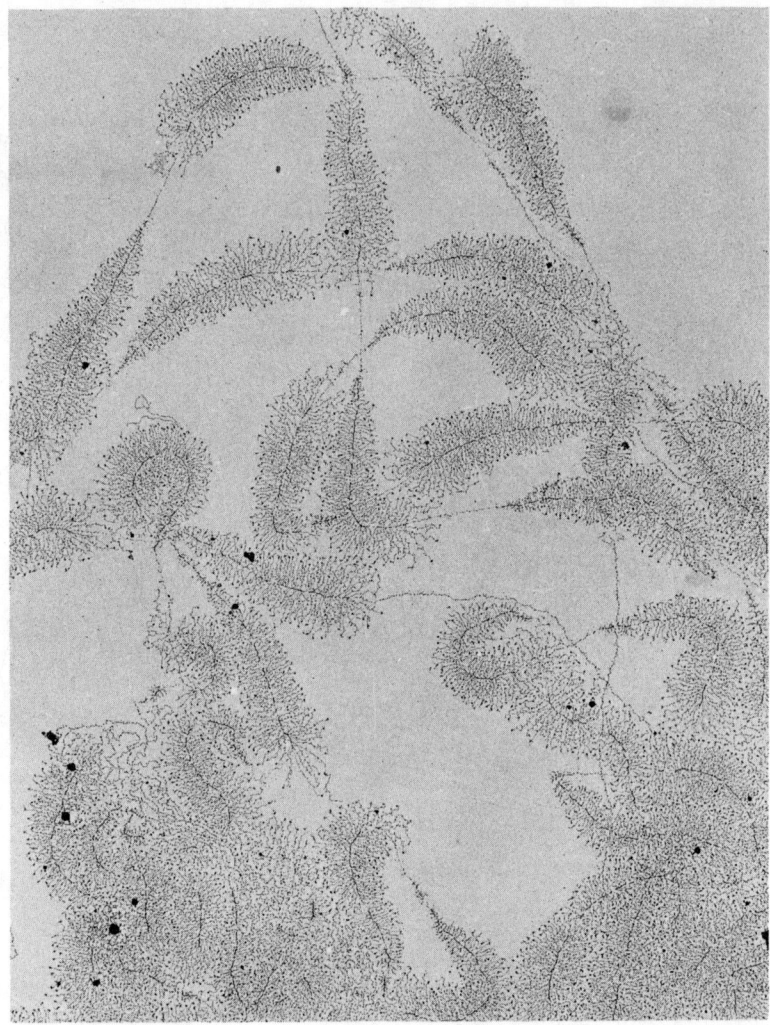

Fig. 2-17   In this electron micrograph, DNA molecules coated with proteins (the long filaments) are transcribed into ribosomal RNA (the branching filaments). There are parts of DNA (bare sections), which appear genetically inactive. The product rRNA molecules get longer as they get completed, so each gene begins wherever the RNA filaments are shortest. Many genes are transcribed simultaneously and also within each gene the transcription process goes on simultaneously for some one hundred rRNA molecules. The DNA is from developing egg cells of the common spotted newt, N. Viridescens; magnification ×13,250.
(Reproduced with permission from: O. L. Miller and B. R. Beatty, J. Cell. Physiol. 74: Sup 1, 225 (1969)). Copyright © Wistar Institute Philadelphia.

How cells differentiate during development into cells of various morphology and functions in a precisely organized pattern to form specialized tissues and organs is still much of a mystery. Of the many interesting systems to study, we give here a glimpse at one, the slime mould, which is particularly convenient as a model system and begins to be understood in some details. The cellular slime mould Dictyostelium discoideum is an amoeba which vegetates nourishing on bacteria. When food gets scanty in the colony, a few cells start releasing periodic pulses of cyclic adenosine monophosphate, cAMP, triggering other cells to propagate the signal. This occurs on a time scale of seconds to minutes allowing for diffusion of the signal, delays in response and in resignaling. The gradient of cAMP elicits a chemotactical response which attracts the cells at velocities of some 5 μm/min towards a few dominant signaling centers, displaying, in the aggregation phase after some 10 hours, characteristic spiral or concentric patterns and steplike coherent movements, which look like periodic waves. Colliding waves annihilate and finally all waves become concentric towards a single center, which has become a nipple-shaped tip. The aggregate forms a columnar conus of some 10 000 cells, which falling over starts migrating as a slug. Fig. 2-18 shows the life cycle of these amoebae. When the slug stops, differentiation occurs in its body: a stalk of dead cells forms which sustains a fruiting body, full of spores. Contrary to early proposals, it appears now that most of the features during aggregation follow from the properties of cAMP and the responses it elicits, an amazingly simple mechanism. The free cells ultimately differentiate into only two types, stalk cells and spore cells. Intense gene activity is found to occur when the tip forms in the aggregate. How the cAMP activity is related to gene activity and differentiation is less understood.

The next steps in understanding would be to see how patterns can emerge in multicellular organisms. These patterns, which are actually organs with diversified functions, are the phenotypes on which the natural selection operates. So it would be extremely interesting to link them to the molecular basis of cell differentiation. This is even less understood at this time. A striking example of the mysteries that are still there is the regeneration of legs in some animals. When the leg of a cockroach or a newt is amputated at any point, at first the wound heals, then dedifferentiated cells accumulate to form a blastema, which ultimately regenerates a complete, functional leg. Regeneration may also occur from the cut surface of the amputated leg: when grafting back a portion of an amputated leg, an intercalary regeneration occurs, to complete the leg as it was originally, Fig. 2-19. However some startling effects occur, if one plays with the length and the orientation of the axis of the grafted material. In this case

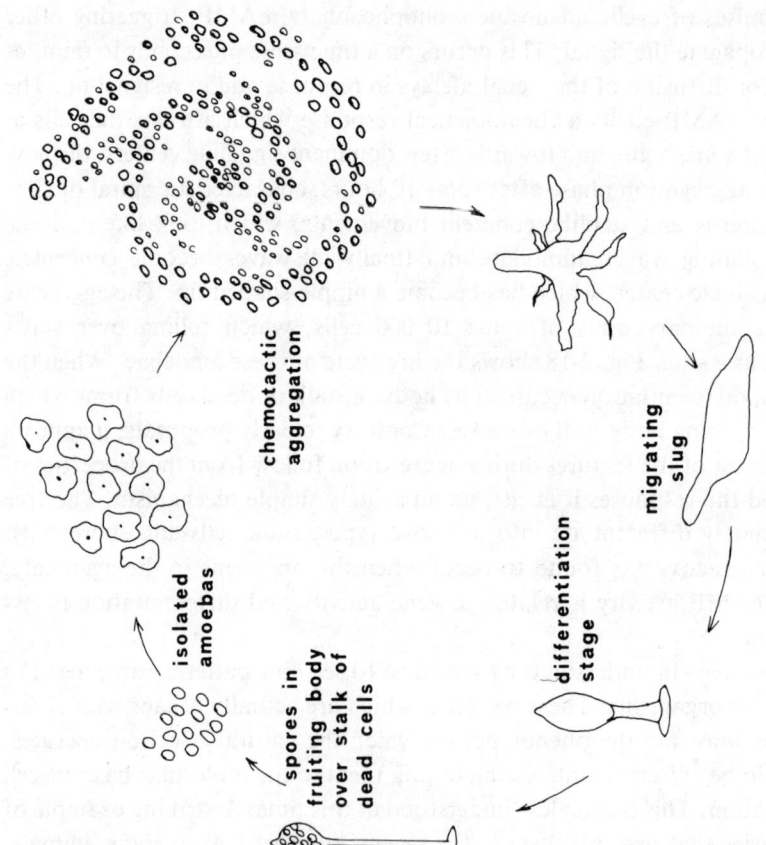

Fig. 2-18   Life cycle of Dictyostelium discoideum (see text).

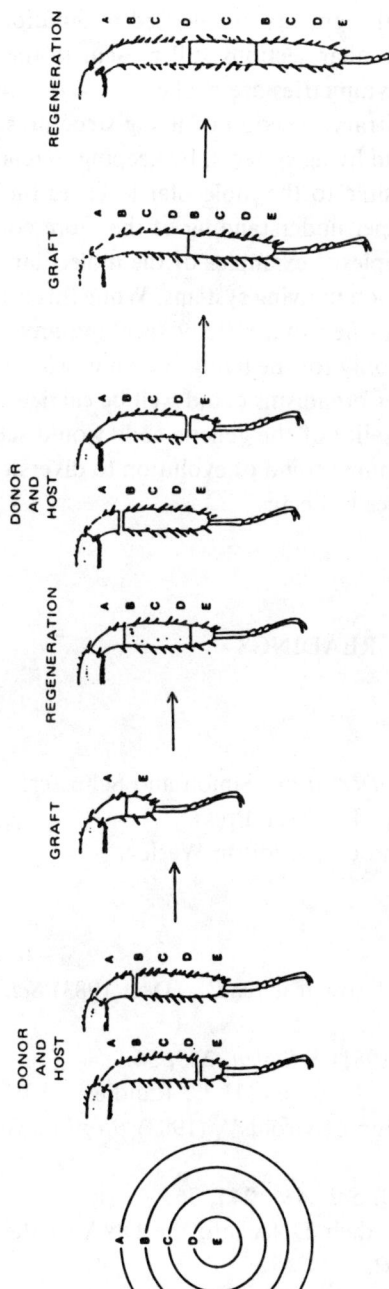

Fig. 2-19 Organization of a leg segment in the cockroach, as observed in the two experiments shown in this illustration, suggests an ordered array. The concentric circles diagram such an array along one axis, from A, the most proximal level (closest to the body), to E, the most distal level. When, in the first experiment, an E-level portion of a tibia is removed from a donor and is grafted onto the A-level portion of the stump of a tibia on a host, local growth at the graft junction produces intercalary regeneration of the missing levels of organization (in this example B, C and D). When, in the second experiment, a proximal-level portion of a tibia from a donor is grafted onto a distal-level portion of a tibia on a host, the confrontation, although it still induces intercalation, gives rise to a reversed intermediate organization level (in this example C between D and B) where bristles grow up rather than down.
(Reproduced with permission from: P. J. Bryant, S. V. Bryant and V. French, Sci. Am. 237, 67 (July 1977)).

monsters can be produced: additional segments of reversed orientation may grow if leg segments are grafted to similar sections still present in the host, supernumeraries appear if left-right symmetries are broken.

We have been showing a quite arbitrary selection of living structures. This reflects the naive attempt to understand living systems, by keeping to relatively simple substructures, as close as possible to the molecular level, in the hope that they will be paradigms for a deeper understanding of the more complex ones. In this spirit we have seen "simplest" examples of the molecular structures which give rise to a specific function in living systems. While this is useful as a first approach, one should be warned that strictly speaking any of the examples would be adequate in detail only for the living system in which it has been studied. Similar function in other organisms could well be carried out in an entirely different way. The universality of the genetic code would seem to permit generalization, while the continuing trend of evolution to diversity and selection permits significant differences to occur.

FURTHER READINGS

General

Judson, H.F. (1979) *The eighth day of creation*, Simon and Schuster.
Luria, S. (1975) *36 lectures in biology*, The MIT Press.
Kimball, J.W. (1978) *Cell biology*, 2nd ed., Addison Wesley.

Specific

Dickerson, R.D. "The DNA helix and how it is read", (Dec. 1983) *Sci. Am.* **249**, 86.
Lake, J.A. "The Ribosome", (Aug. 1981) *Sci. Am.* **245**, 56.
Riesner, D., Steger, G., Schumacher, J., Cross, H.J., Randles, J.W. and Sänger, H.L. "Structure and function of viroids", (1983) *Biophys. Struct. Mech.* **9**, 145.
Novick, R.P. "Plasmids", (Dec. 1980) *Sci. Am.* **243**, 76.
Campbell, A.M. "How viruses insert their DNA into the DNA of the host cell", (Dec. 1976) *Sci. Am.* **235**, 102.
"The molecular basis of life", (1966) *Readings from Scientific American* Freeman Publ. Co.

"The living cell", (1966) *Readings from Scientific American*, Freeman Publ. Co.

Lodish, H.L. and Rothman, J.E. "The assembly of cell membranes", (Jan. 1979) *Sci. Am.* **240**, 38.

Chance, B., Mueller, P., De Vault, D. and Powers, L. (Oct. 1980) "Biological membranes", *Physics Today* p. 32–38.

Unwin, N. and Henderson, R. "The structure of proteins in biological membranes", (Feb. 1984) *Sci. Am.* **250**, 56.

Perutz, M.F. "Hemoglobin structure and respiratory transport", (Dec. 1978) *Sci. Am.* **239**, 68.

Berg, H.C. "How bacteria swim", (Aug. 1975) *Sci. Am.* **233**, 36.

Bauer, W.R., Crick, F.H.C. and White, J.H. "Supercoiled DNA", (July 1980) *Sci. Am.* **243**, 100.

Porter, K.R. and Tucker, J.B. "The ground substance of the living cell", (March 1981) *Sci. Am.* **244**, 41.

Tyler Bonner, J. "Chemical signals of social amoebae", (Apr. 1983) *Sci. Am.* **248**, 106.

Bryant, P.J., Bryant, S.V. and French, V. "Biological regeneration and pattern formation", (July 1977) *Sci. Am.* **237**, 67.

Wolpert, L. "Pattern formation in biological development", (Oct. 1978) *Sci. Am.* **239**, 124.

CHAPTER 3

LIVING SYSTEMS AND THERMODYNAMICS

*We can use physical models to
disprove any claim that the known
laws of physics are not sufficient to
describe the phenomena which specify
a living object.*

Manfred Eigen

The high degree of order displayed by living systems in space and time makes one often wonder about their compatibility with the laws of thermodynamics, specifically with the second law. At first glance the tendency of living systems to increase their internal order in the course of their differentiation, growth, development, may appear to be at odds with the tendency of most condensed matter to proceed towards states in which the entropy of the system is maximized. On the other hand one may wonder what the order seen in living systems has to do with the order displayed by systems of condensed matter undergoing order-disorder phase transitions. A few straightforward comments serve to clear up this point. The requirement of the second law that there be an increase of entropy applies to isolated systems, which, when the maximum entropy is reached, are in an equilibrium state. If the system is not isolated, we may, for instance, extract heat from it and cool it and observe phase transitions to equilibrium states, which minimize the free energy $F = U - TS$ by trading a decrease in entropy, S, with a larger decrease in internal energy, U, at the temperature T. The system below some critical temperature may display order at equilibrium.

A living system is not isolated and is not in equilibrium. It is rather an open system, which can exchange matter and energy with the environment. It is in a stationary state which is not at equilibrium, but which must display stability for times, which are short with respect to a lifetime and long with respect to the characteristic times of the internal processes of the system. It can evolve continuously to other stationary states of slightly different structure and function in times comparable to its lifetime.

Here we shall see what the extension of thermodynamics to stationary states which are not at equilibrium can tell us and the limits of such an approach. Our aim is simply to answer the basic questions outlined above about the compatibility of order and dissipation.

Let us consider an open system, which can exchange heat, work, matter and, therefore, entropy with its surroundings. If we make the assumption that all of the thermodynamic parameters, and especially the entropy, are functions of only those variables of which they were functions at equilibrium, then we can divide the entropy change dS into that flowing from the outside d_eS, with plus sign when flowing into the system, and that produced in the system d_iS, so that

$$dS = d_eS + d_iS \tag{1}$$

for the system under consideration. Now we accept the evidence that the biochemical processes occurring inside the system are typically irreversible as is any chemical reaction which is not strictly at equilibrium. Indeed, as is the case for any chemical reaction, they are trying to proceed towards equilibrium, but in the steady living state they are frequently not permitted to do so because of being linked with other reactions or processes such as the transport of materials across boundaries. The irreversibility of these chemical reactions requires that $d_iS > 0$, in accord with the second principle of thermodynamics. Since at a stationary state $dS = 0$, by definition we see that $d_eS < 0$, that is the system is exporting its internal entropy production into the rest of the universe. In a successful attempt to maintain its stationary state the system is helping to increase the entropy of the universe. This does not show us how order comes about in a living system, but simply that it is not in contradiction with dissipation.

To make use of the straightforward argument we presented above, we had to rely on a rather strong assumption, which implies the following: that we can find an intermediate scale of time and volumes which is small enough so that the external fields and the thermodynamic parameters themselves do not vary appreciably on such a scale and at the same time it is large enough with respect to the microscopic features of the system that it is meaningful to take thermodynamic averages, which are needed in turn to define locally such proper thermodynamic parameters as temperature and entropy. This hypothesis, that the system is locally well behaved, is by no means trivial nor generally obeyed in non-equilibrium systems. For a thorough discussion of the limits of this assumption in the classical case of an ideal gas, one of the few which is tractable from first principles, we refer the reader to the book by Kreutzer.

The local approach is adequate at least for many problems, such as the transport of both charged and uncharged molecules through membranes and biochemical reactions which proceed close to equilibrium. This is actually the case for many processes of interest, because conditions far from equilibrium are often displayed in connection with the control of metabolic pathways and therefore are a property of only a few key steps. If we accept this, then we can extrapolate the Gibbs formula, which is valid for open systems at equilibrium, to systems not at equilibrium

$$T dS = dU + P dV - \Sigma_k \mu_k dN_k \qquad (2)$$

where P and V are the pressure and volume of the system, N_k is the molar amount of the kth component and we have introduced the chemical potential μ_k of kth component. From this equation it is obvious that

$$\mu_k = \left(\frac{\partial U}{\partial N_k}\right)_{S, V, N_{k'} \neq k} \qquad (3)$$

The chemical potential represents the change in internal energy of the system, when one mole of the kth component is added to the system while keeping the entropy, the volume V and the molar amounts N_k, of the other components constant. For a system at constant temperature and pressure the chemical potential takes the form

$$\mu_k = \left(\frac{\partial G}{\partial N_k}\right)_{T, P, N_{k'} \neq k} \qquad (4)$$

where G is the Gibb's free energy, $G = U - TS + PV$. For ideal solutions, at constant pressure we can make the approximation that

$$\mu_k = \gamma_k(P, T) + RT\ln n_k \tag{5}$$

where $n_k = N_k/\Sigma_k N_k$ is now the molar fraction of the kth component, R is the gas constant and one disregards the possible influence on μ_k from the presence of the other components k'; if we consider that in general a chemical reaction will involve the participation of several components of our system, for the αth reactions we can define an affinity A_α as

$$A_\alpha = -\Sigma_k \nu_{\alpha k} \mu_k \tag{6}$$

where $\nu_{\alpha k}$ is the stoichiometric coefficient with which the kth component enters the αth reaction, the sign being positive for products and negative for reactants. Taking the expression for μ_k in ideal solutions, we obtain that

$$A_\alpha(P, T, n_k) = RT\ln [\, K_\alpha(P, T)\Pi_k n_k^{-\nu_{\alpha k}} \,] \tag{7}$$

where in the equilibrium constant $K_\alpha(P, T)$, we have lumped all the concentration independent $\gamma_k(P, T)$ terms as $\ln K_\alpha(P, T) = -\Sigma_k \nu_{\alpha k}\gamma_k(P, T)$. We now introduce a term for the extent of completion of the αth reaction, ξ_α such that

$$d\xi_\alpha = -\frac{dN_k}{\nu_{\alpha k}}. \tag{8}$$

This function is obviously independent of the index k since

$$\frac{dN_k}{\nu_{\alpha k}} = \frac{dN_{k'}}{\nu_{\alpha k'}}. \tag{9}$$

The affinity of the reaction can now be written as

$$A_\alpha = -\left(\frac{\partial G}{\partial \xi_\alpha}\right)_{P, T} \tag{10}$$

which shows its physical meaning more clearly. As G must be a minimum at equlibrium, when P and T are kept constant, we require the chemical reaction to proceed in such a way as to approach the state in which

$$\left(\frac{\partial G}{\partial \xi_\alpha}\right)_{P,T} = 0. \tag{11}$$

Therefore $A_\alpha = 0$ at equilibrium and we get the familiar expression for the chemical equilibrium of the αth reaction

$$K_\alpha(P, T) = \Pi_k n_k^{\nu_{\alpha k}} \tag{12}$$

If now we drive the system out of equilibrium, we can write the entropy production due to the chemical reaction in the form

$$\frac{d_i S}{dt} = \Sigma_\alpha \frac{A_\alpha}{T} \frac{d\xi_\alpha}{dt} > 0 \tag{13}$$

where we have used the Gibbs equation, eq. (2), and the definition of $d_i S$ given in equation (1). This equation indicates that if we have only one chemical reaction it is necessary that A and $d\xi/dt$ have the same sign and therefore the direction towards which a reaction goes is determined by the thermodynamic variable $A(P, T, \{n_k\})$.

This rather abstract formulation, which is not generally used in chemistry and biochemistry, has the advantage that it allows us to make a few general considerations about living systems, in which we find biochemical reactions to be interlinked and often coupled to other dissipative processes such as transport through membranes, diffusion etc. We shall make use of this formalism, in order to illustrate these features with a few specific examples.

As a first straightforward example let us consider a set of linked reactions, for instance a chain of consecutive reactions, $C_k \rightarrow C_{k+1}$, for a number of different chemical species C_k. Given the sign of the total affinity $A = \Sigma_\alpha A_\alpha$, we can predict in which direction the system evolves globally, but in detail, since we only require that

$$\Sigma_\alpha A_\alpha \frac{d\xi_\alpha}{dt} > 0, \tag{14}$$

we may well have

$$A_\beta \frac{d\xi_\beta}{dt} < 0 \tag{15}$$

for some β. One or more of the reactions may go backwards, i.e. against their affinity. This is a feature quite similar to one displayed by living systems, where we see chemicals being transported against concentration gradients, and reactions being forced to accumulate rather than release free energy, etc.

Before going to another simple example, let us make an interesting observation about $d\xi/dt$ and A, which is valid close to equilibrium. Let us consider the reaction $B + C \overset{k_+}{\underset{k_-}{\rightleftharpoons}} D$. The kinetic constants k_+ and k_- give the velocities of reaction in the forward and backward directions, v_+ and v_- respectively, as functions of the concentrations $[B]$, $[C]$ and $[D]$ of the reactants; $v_+ = k_+[B][C]$, $v_- = k_-[D]$. At equilibrium $d\xi/dt = v_+ - v_- = 0$ and $k_-[D] = k_+[B][C]$. From equation (12), valid at equilibrium, we obtain

$$K(P, T) = \frac{k_+}{k_-}. \tag{16}$$

When not at equilibrium $d\xi/dt \neq 0$ and of course

$$\frac{d\xi}{dt} = k_+[B][C] - k_-[D] \tag{17}$$

Introducing the affinity A through its expression in terms of $K(P, T)$ eq. (7), we easily obtain

$$\frac{d\xi}{dt} = v_+(1 - e^{-A/RT}). \tag{18}$$

Equation (7) in this case is written as

$$A = RT \ln K(P, T)\frac{[B][C]}{[D]}. \tag{19}$$

Now if we assume that $[B]$, $[C]$, and $[D]$ are very close to their equilibrium values, then $A << RT$ and we can approximate equation (18) as

$$\frac{d\xi}{dt} \simeq \frac{v_{+,\text{eq}}}{RT} A \tag{20}$$

where $v_{+,\text{eq}} = v_{-,\text{eq}}$ is the value of the velocity of reaction at equilibrium.

Therefore, for systems which are close to equilibrium we can express the time derivative of the extent of completion of the reaction as a linear function of the affinity as

$$\frac{d\xi}{dt} = L\frac{A}{T} \tag{21}$$

where L is a phenomenological coefficient which does not depend on ξ but only on P and T, $L = L(P, T)$.

For a set of coupled reactions we have the obvious extension

$$\frac{d\xi_\alpha}{dt} = \Sigma_\beta L_{\alpha\beta}\frac{A_\beta}{T} \tag{22}$$

where the terms $L_{\alpha\beta}$ describe the coupling between different reactions. For instance for two coupled reactions we have the set of coupled equations

$$\frac{d\xi_\alpha}{dt} = L_{\alpha\alpha}\frac{A_\alpha}{T} + L_{\alpha\beta}\frac{A_\beta}{T}$$

$$\frac{d\xi_\beta}{dt} = L_{\beta\alpha}\frac{A_\alpha}{T} + L_{\beta\beta}\frac{A_\beta}{T} . \tag{23}$$

For the off-diagonal terms $L_{\alpha\beta}$, it was shown by Onsager that, in the linear region close to equilibrium, these relationships display the remarkable property that

$$L_{\alpha\beta} = L_{\beta\alpha} . \tag{24}$$

This formalism can be extended to include all irreversible processes occurring at steady states close to equilibrium, such as diffusion against either concentration or electrochemical gradients, osmosis, etc., provided the basic assumptions of the local formulation for systems close to equilibrium are fulfilled. Specifically for an open system, which can exchange materials across its boundary, one can consider the flux,

$$\frac{d_e n_m}{dt} ,$$

of the mth component from the external environment. In the linear region, close to equilibrium this will be given by

$$\frac{d_e n_m}{dt} = L_{mm} \frac{A_{mm}}{T} + L_{mp} \frac{A_{mp}}{T} + \ldots \qquad (25)$$

where now A_{mm} represents the affinity for the transport of the mth component across the boundary of the system and the off-diagonal terms A_{mp} describe the coupling of this transport process with the transport of other components p, or with other processes.

Let us consider now the following simple example: the transport of chemicals in and out of an open system, inside which a chemical reaction occurs involving some of the same chemicals. The system and the environment are kept at constant temperature and pressure. The component m enters the system where it is converted into component p, which can leave the system. The transport of m across the boundary of the system is coupled to that of another component s, which however does not participate to any chemical reaction inside the system. In the linear approximation we obtain

$$\frac{d_e n_m}{dt} = L_{mm} \frac{A_m}{T} + L_{ms} \frac{A_s}{T}$$

$$\frac{d_e n_s}{dt} = L_{sm} \frac{A_m}{T} + L_{ss} \frac{A_s}{T}$$

$$\frac{d_e n_p}{dt} = L_{pp} \frac{A_p}{T} \qquad (26)$$

$$\frac{d\xi}{dt} = L \frac{A}{T}$$

where $d\xi/dt$ and A are respectively the velocity and the affinity of the reaction $m \rightarrow p$ and the other symbols are obvious. At steady state we have that the net change with time of the molar amount of each component must be zero, that is

$$\frac{dn_x}{dt} = \frac{d_e n_x}{dt} + \frac{d_i n_x}{dt} = 0 \qquad (27)$$

where we have divided the change in the molar amount of the xth component dn_x into its contribution $d_e n_x$ from the environment and its contribution $d_i n_x$ from processes internal to the system. Of course in our case we have

$$-\frac{d_i n_m}{dt} = \frac{d_i n_p}{dt} = \frac{d\xi}{dt} \,.$$

(28)

For the steady state conditions, using eqs. (27) and (28) we obtain

$$\frac{dn_m}{dt} = \frac{d_e n_m}{dt} - \frac{d\xi}{dt} = 0$$

$$\frac{dn_p}{dt} = \frac{d_e n_p}{dt} + \frac{d\xi}{dt} = 0$$

(29)

$$\frac{dn_s}{dt} = \frac{d_e n_s}{dt} = 0 \,.$$

In the linear region, close to equilibrium, the Onsager relations eq. (24) apply. Solving for the affinities we find

$$A_m = \frac{T}{L_{mm} - \dfrac{L^2_{ms}}{L_{ss}}} \frac{d\xi}{dt}$$

$$A_p = -\frac{T}{L_{pp}} \frac{d\xi}{dt}$$

(30)

$$A_s = \frac{-T\dfrac{L_{ms}}{L_{ss}}}{L_{mm} - \dfrac{L^2_{ms}}{L_{ss}}} \frac{d\xi}{dt}$$

What is this telling us? We have assumed that the environment in which the open system is embedded is an unlimited source of component m and an unlimited sink for component p. They both cross the boundary of the system according to intrinsic properties of the boundary described by the coefficients L_{mm}, L_{ss}, L_{pp}, and L_{ms}, which are assumed to be independent of the fluxes and affinities of the processes we are considering. They depend only on the thermodynamic parameters which characterize the system and its environment, in our case the temperature, T, and pressure, P, which are both held constant. We have calculated the affinities A_m, A_p and A_s of the three processes in our system at steady state as functions of temperature, the $L_{\alpha\beta}$ coefficients, and the rate, $d\xi/dt$, of the reaction $m \rightarrow p$. If the transport of s were not coupled to that of m, i.e. $L_{ms} = 0$, then at steady state the affinity, A_s, of this process would be $A_s = 0$ and it would be at equilibrium. Assuming that for transport processes between ideal solutions $K(P, T) = 1$, this would mean that $[s]_{in} = [s]_{out}$. In contrast, by coupling the transport of s to that of m, a steady state is reached at which $A_s \neq 0$. Since for ideal solutions

$$A_s = RT \ln(\frac{[s]_{out}}{[s]_{in}}) \tag{31}$$

a concentration difference or gradient in s is maintained, which depends on the speed of the $m \rightarrow p$ reaction. Notice that even if $L_{ms} = 0$, the steady state values of A_m and A_p will be non-zero because these two processes will still be linked. Once more this is a typical feature of living systems, which can be modeled by near-equilibrium thermodynamics. We have promoted gradients of concentration of some components across a boundary, quite different from those expected by considering only the diffusion of that component. This has been achieved with the expenditure of chemical energy by the chemical reaction which is coupled to the transport processes under consideration.

Now it is interesting to note that, in the linear region, because of the Onsager reciprocity relations, eq. (24), we can predict what steady state will be attained by the system under the given set of constraints and also whether or not it will be stable against small perturbations. We have already seen in eq. (13) that the affinities A_α and the velocities $d\xi_\alpha/dt$ of reaction appear in the combination $(A_\alpha/T)(d\xi_\alpha/dt)$ in the expression for the production of entropy, where the "flux" $d\xi/dt$ is driven by "force" A_α/T. This is a general feature of the irreversible processes and we can talk about "conjugated" forces X_α and fluxes J_α. Until now we have $X_\alpha \equiv A_\alpha/T$ and $J_\alpha \equiv d\xi_\alpha/dt$. In general the conjugated pairs X_α, J_α may describe any dissipative transport process such as the conduction of electricity, of heat, the diffusion of chemicals, etc. Also we have shown that, for chemical reactions in the linear region, we can express the fluxes as linear functions of the forces through the coefficients L. Under proper conditions this is also true for these other dissipative processes. Therefore, for two coupled irreversible processes close to equilibrium we can write as in equations (13) and (23)

$$\frac{d_i S}{dt} = J_\alpha X_\alpha + J_\beta X_\beta \tag{32}$$

and

$$J_\alpha = L_{\alpha\alpha} X_\alpha + L_{\alpha\beta} X_\beta \tag{33}$$

$$J_\beta = L_{\alpha\beta} X_\alpha + L_{\beta\beta} X_\beta$$

with the Onsager relations expressed by $L_{\alpha\beta} = L_{\beta\alpha}$. It follows that

$$\frac{d_i S}{dt} = L_{\alpha\alpha} X_\alpha^2 + 2 L_{\alpha\beta} X_\alpha X_\beta + L_{\beta\beta} X_\beta^2 . \tag{34}$$

Since the quadratic form $d_i S/dt$ must be positive definite, because the second law of thermodynamics requires that $d_i S/dt \geq 0$, we get that $L_{\alpha\alpha}, L_{\beta\beta} > 0$ and $|L_{\alpha\beta}|^2 < 4 L_{\alpha\alpha} L_{\beta\beta}$. This says that the "proper" coefficient must be positive, like thermal conductivity, diffusion constants etc., while the coupling coefficient $L_{\alpha\beta}$ may show any sign.

Now, back to the stability of steady states, let us assume, as an external constraint, that $J_\beta = 0$. What is the behavior of $d_i S/dt$? Which of all the possible steady states is attained and is in some sense stable? If we try to change the conjugated driving force X_β, we obtain

$$\frac{\partial}{\partial X_\beta} \frac{d_i S}{dt} = 2(L_{\alpha\beta} X_\alpha + L_{\beta\beta} X_\beta) = 2 J_\beta = 0 . \tag{35}$$

As $J_\beta = 0$ is our constraint, this means that the steady state attained is that giving an extremum for the entropy production in respect to a variation of the driving force X_β, when X_α is kept constant. As

$$\frac{\partial^2}{\partial X_\beta^2} \frac{d_i S}{dt} = 2 L_{\beta\beta} > 0 \tag{36}$$

we see that the extremum is a minimum. So the steady state is the one which is characterized by a minimum in the entropy production under the specified constraints. This is valid only in the linear region because we had to use $L_{\alpha\beta} = L_{\beta\alpha}$ to get the definite signs of $L_{\alpha\alpha}$ and $L_{\beta\beta}$.

It is instructive to investigate the stability by a somewhat different method. Let us focus on two coupled chemical reactions. Let the system be closed with respect to the exchange of matter, so that ξ_1 and ξ_2 are the only variables, P and T be constant and $dL_{\alpha\beta}/dt = 0$. The affinities A_α will be functions of the form $A_\alpha = A_\alpha(P, T, \xi_1, \xi_2)$. Then from the linear relations we obtain

$$\frac{d}{dt}\frac{d_iS}{dt} = \frac{2}{T}\left(\frac{d\xi_1}{dt}\frac{dA_1}{dt} + \frac{d\xi_2}{dt}\frac{dA_2}{dt}\right)$$

$$= \frac{2}{T}\left\{\frac{d\xi_1}{dt}\left[\left(\frac{\partial A_1}{\partial\xi_1}\right)_{P,T}\frac{d\xi_1}{dt} + \left(\frac{\partial A_1}{\partial\xi_2}\right)_{P,T}\frac{d\xi_2}{dt}\right] + \right.$$

$$\left. \frac{d\xi_2}{dt}\left[\left(\frac{\partial A_2}{\partial\xi_1}\right)_{P,T}\frac{d\xi_1}{dt} + \left(\frac{\partial A_2}{\partial\xi_1}\right)_{P,T}\frac{d\xi_2}{dt}\right]\right\} \qquad (37)$$

we recall that

$$\left(\frac{\partial A_2}{\partial\xi_1}\right)_{P,T} = -\left(\frac{\partial^2 G}{\partial\xi_1\partial\xi_2}\right)_{P,T} = -\left(\frac{\partial}{\partial\xi_2}\frac{\partial G}{\partial\xi_1}\right)_{P,T} = \left(\frac{\partial A_1}{\partial\xi_2}\right)_{P,T}. \qquad (38)$$

Therefore

$$\frac{d}{dt}\frac{d_iS}{dt} = \frac{2}{T}\left\{\left(\frac{\partial A_1}{\partial\xi_1}\right)_{P,T}\left(\frac{d\xi_1}{dt}\right)^2 + 2\left(\frac{\partial A_1}{\partial\xi_2}\right)_{P,T}\frac{d\xi_1}{dt}\frac{d\xi_2}{dt} + \right.$$

$$\left. \left(\frac{\partial A_2}{\partial\xi_2}\right)_{P,T}\left(\frac{d\xi_2}{dt}\right)^2\right\} \qquad (39)$$

Close to equilibrium

$$A \simeq \left(\frac{\partial A}{\partial\xi}\right)_{eq}(\xi - \xi_{eq}). \qquad (40)$$

Since for any fluctuation taking the system away from equilibrium we must have $\Delta_i S < 0$, because S is a maximum at equilibrium, we have that

$$\Delta_i S = \int_{\xi_{eq}}^{\xi} d_i S = \int_{\xi_{eq}}^{\xi} \frac{A}{T} d\xi \simeq \frac{1}{2T} \left(\frac{\partial A}{\partial \xi}\right)_{eq} (\xi - \xi_{eq})^2 \tag{41}$$

For more than one reaction of course we obtain

$$\Delta_i S = \frac{1}{2T} \Sigma_{\alpha\beta} \left(\frac{\partial A_\alpha}{\partial \xi_\beta}\right)_{eq} \Delta\xi_\alpha \Delta\xi_\beta . \tag{42}$$

So the coefficients

$$\left(\frac{\partial A_\alpha}{\partial \xi_\beta}\right)_{eq}$$

must render the quadratic form negative definite. Similarly, the quadratic form $(d/dt)(d_i S/dt)$ must be negative definite since it has the same set of coefficients. At the same time $d_i S/dt$ is positive definite. Therefore as the dissipation $d_i S/dt$ is always positive and cannot increase, it will eventually go to a minimum, the stable steady state.

It should be quite clear now how these minimum properties are directly related to the second law of thermodynamics, which allows this extension of thermodynamics to states not at but close to equilibrium. This result can be easily extended to N linked reactions. Let the system be closed. We consider the deviation from equilibrium for each reactant

$$x_i = [x_i] - [x_i]_{eq} \tag{43}$$

where $[x_i]_{eq}$ are given by the equilibrium conditions

$$A = 0, \qquad \frac{dx_i}{dt} = 0 . \tag{44}$$

Because of conservation of mass, $\Sigma_i [x_i] = $ constant, and the other constraints, when the system deviates from equilibrium we have in general $n < N - 1$ independent equations

$$\frac{\mathrm{d}x_i}{\mathrm{d}t} = f_i(x_1, x_2 \ldots x_n) \qquad i = 1, \ldots, n \tag{45}$$

where $f_i(x_1, \ldots, x_{n-1})$ is a nonlinear function of the x_i. If we are close enough to equilibrium, all of the kinetic equations can be linearized and the Onsager relations are also valid with a proper choice of the fluxes and conjugated forces.

It can be shown that, when such a linearization is performed, the general solution of eq. (45) is a set of $x_i(t)$, which always approaches zero in time as a linear combination of exponentials with at most n different time constants. The rather formal mathematical proof of this assertion resides not only on the linearization, consequent to the assumption of linear forces versus fluxes relations, but also on the symmetry of these relations due to the Onsager identities. It follows that, quite in general, no steady oscillation or repetitive time pattern is allowed. Now we replace the parameters for an equilibrium state as considered in the above analysis with those for a steady state close to equilibrium. We obtain identical results, as should be evident from the discussion above. Specifically it is obvious that all what must be done is to substitute the $[x]_{eq}$ with $[x]_{ss}$, where the subscript ss indicates steady state values. Therefore if we disturb the system from a steady state close to equilibrium, it will go back to that steady state with exponential relaxation. Again we see that no repetitive pattern in time is permitted. As this behavior is ultimately linked to the symmetry properties of the kinetic equations which are due to the validity of the Onsager relations, it is clear that the absence of oscillatory behavior, or more generally of patterns in space and time, is actually prohibited by the fact that the system is in the linear region close to equilibrium.

The above discussion has shown that irreversible thermodynamics in the linear region close to equilibrium is a convenient framework in which to describe many of the physico-chemical processes necessary for living systems. However, we have explicitly shown that under these conditions no oscillatory behavior can be produced. It may be surmised that no structure or pattern in space can be generated either. Therefore, something is missing since living systems show just the opposite properties. They show cycles in behavior, in chemical composition and in the activity of cells; in addition cells of definite size and shape organize to form organs and organisms. Even within cells there are compartments. These patterns in space and time develop on all scales, from submolecular in space and seconds in time to full organisms and genes. How does this occur? For a long time this was regarded as "the" secret of life.

Recently, while by no means solved in a general way, the problem seems to have been attacked with a fruitful approach.

The points to be considered are the following: i) as irreversible processes, reactions can be driven far from equilibrium with comparative ease; ii) far from equilibrium the dynamic equations describing the space-time evolution of the concentrations of the various components of the system show non linearities; iii) sets of nonlinear, coupled differential equations may have solutions which oscillate in time or have structure in space for special sets of parameters. Let us discuss these points.

If we consider a typical irreversible process such as the flow of a viscous fluid in a channel, it shows a laminar Pouseille flow. The fluid velocity increases smoothly and regularly from zero at the walls to its maximum value at the central region of the channel. This is the behavior which corresponds to our linear, close to equilibrium condition. It is common experience that most fluids show this linear behavior for quite a wide range of average velocities with respect to the channel. Only at comparatively high velocities, in correspondence with the so called Reynolds number, does the linear behavior break down producing turbulence. Under these conditions the fluid velocity field shows a chaotic behavior in space and time. For chemical reactions we have already discussed the fact that linearity and the close to equilibrium condition are obtained if the affinity A is very close to zero, its equilibrium value, as given by $A \ll RT$. In general from the definition of A and the expression for the chemical potential for ideal systems, equations (6) and (5) respectively we obtain equation (7) which relates the affinity, A, to the concentrations of the various components and to the equilibrium constant $K(T)$. It is evident from equation (7) and from equation (12), the expression of $K(T)$ in terms of the equilibrium values of the reactant concentrations, that $A \ll RT$ only when the various reactant concentrations are very close to their equilibrium values. This is not a common situation for chemical reactions. On the contrary it is comparatively easy to have $n_k \gg n_{k,\text{eq}}$ or $n_k \ll n_{k,\text{eq}}$. This ease with which the non-linear region is reached seems to be an intrinsic property of chemical reactions.

In order to investigate the behavior of chemical systems which are far from equilibrium when $A \gg RT$ and one expects non-linearity, one may try to extend the thermodynamics of irreversible processes. Keeping the local formulation one makes the following assumptions: it is still meaningful to define entropy locally, that such an entropy is a function only of those variables upon which it depends at equilibrium and that the Gibbs formula, eq. (2), applies. In analogy to the principle of minimum entropy production, which we have seen to be valid close to equilibrium, there has been a search for a general principle of evolution, which would predict the steady states to which an open system which is far from equilibrium would spontaneously evolve when perturbed under specified constraints. The usefulness of such an approach has been questioned primarily because it appears to be more difficult to investigate the stability of steady states against perturbations, which of course include spontaneous fluctuations. The problem is that even if we take as valid the local formulation, it is in general not possible to construct functions of the entropy and of the production of entropy which show a definite behavior when the system evolves toward a stable steady-state. Again we may notice the difference from the situation close to equilibrium where the sign of the entropy production and of its time derivative ensure evolution towards the stable steady state. In some special cases of systems far from equilibrium it is possible to construct a function called "excess entropy production" which shows definite behavior in the vicinity of a stable steady state, but the conditions given by the related evolution criteria are only sufficient and not necessary. Modifications of the local formulation have been pursued to allow for functions which reduce to ordinary entropy when the system approaches or reaches equilibrium but which generalize the concept of entropy far from equilibrium by allowing a dependence on variables other than those of which entropy is a function at equilibrium, such as the velocity of reaction. These approaches seem to permit the introduction of criteria of evolution of more general validity. We will not enter this discussion here and refer the interested reader to the books and reviews by Nicolis and Prigogine, by Keizer and by Kreutzer.

In order to find how an open system can spontaneously evolve to a quasi-steady state which is far from equilibrium and which displays ordered structures in space and time it appears more fruitful to investigate the behavior of specific sets of kinetic equations, which often describe sets of coupled reactions of actual biochemical interest. Here again we need the local formulation to be valid, because we want to describe the system using kinetic constants, diffusion coefficients, etc. with the same meaning and values we observe close to equilibrium. We do this not simply because it would be very inconvenient to do without the local formulation, but because it has been found to be appropriate in most relevant situations. In experiments in chemical kinetics, for example, the kinetic constants which describe systems far from equilibrium appear to keep their same values as the systems approach their equilibrium states. In general these sets of kinetic equations, a few of them quite famous, have been developed in connection with a variety of different problems. Historically remarkable examples are the Lotka-Volterra equations, the Turing machine, and the so called Brusselator; for details see the book by Nicolis and Prigogine. The intrinsic non-linearity of these sets of coupled equations, when they are driven far from their trivial equilibrium solutions by the choice of constraints, produce solutions which evolve with oscillations in time and non-uniformities in space. The stabilities of these solutions are by no means easy to investigate mathematically and considerable effort is currently being spent towards developing more or less general methods to approach this problem. Here we will give an elementary discussion of one such method and present a few specific examples.

Virtually any metabolic pathway with catalytic steps and linked reactions shows intrinsic non-linearity in the differential equations by which it is described. Let us consider a model described by Lotka and Volterra to describe a prey-predator system. The prey, X, feeds on an unlimited food source, F, and therefore reproduces at a rate $k_1[F]$. The predator, Y, feeds only on the prey. The prey disappears only because of the interaction with the predator. The predator reproduces proportional to the disappearance of the prey and dies of natural causes at a specified rate k_3 giving rise to a population, D, of deads. The system is open as we keep the food concentration $[F]$ constant and also constrain $[D]$ to be constant. The kinetic equations for the populations $[X]$ and $[Y]$ of X and Y are

$$\frac{d[X]}{dt} = k_1[F][X] - k_2[X][Y]$$

$$\frac{d[Y]}{dt} = k_2[X][Y] - k_3[Y]$$

(46)

This system is in a steady state if $d[X]/dt = d[Y]/dt = 0$. Apart from the trivial solution $[X] = [Y] = 0$, this is achieved when $[X]_{ss} = k_3/k_2$ and $[Y]_{ss} = k_1[F]/k_2$.

Is this steady state stable with respect to small fluctuations? That is to say, will fluctuations in $[X]$ and $[Y]$ be damped out so that these quantities relax back to $[X]_{ss}$ and $[Y]_{ss}$? We can attempt to see what happens to the deviations from steady state, $x = [X] - [X]_{ss}$ and $y = [Y] - [Y]_{ss}$, in much the same way as we did for deviations from equilibrium and steady states close to equilibrium. Substituting for x and y in the above equation we get

$$\frac{\mathrm{d}x}{\mathrm{d}t} = k_1[F]x - k_2\{[Y]_{ss}x + [X]_{ss}y + xy\}$$

$$\frac{\mathrm{d}y}{\mathrm{d}t} = k_2\{[Y]_{ss}x + [X]_{ss}y + xy\} - k_3 y \tag{47}$$

Inserting the values for $[X]_{ss}$ and $[Y]_{ss}$ and disregarding the terms xy to keep to first order in x and y we obtain

$$\frac{\mathrm{d}x}{\mathrm{d}t} = -k_3 y$$

$$\frac{\mathrm{d}y}{\mathrm{d}t} = +k_1[F]x \tag{48}$$

If we look for solutions of the kind $x = x_0 e^{\lambda t}$, $y = y_0 e^{\lambda t}$ we obtain

$$\lambda^2 = -k_2 k_1[F]$$

which gives a purely imaginary number since k_2, k_1, and $[F]$ are intrinsically positive quantities. The fluctuations around the steady state are not damped, but rather oscillate at a characteristic frequency $\gamma = 2\pi k_1 k_2[F]$ which depends on the reproductive properties of the prey, k_1, on the ability of predator to catch the prey, k_2, and on the food supply $[F]$.

This same model can be easily put into the language of chemistry by writing the following coupled reactions:

$$F + X \underset{k_{-1}}{\overset{k_1}{\rightleftharpoons}} 2X$$

$$X + Y \underset{k_{-2}}{\overset{k_2}{\rightleftharpoons}} 2Y \tag{49}$$

$$Y \underset{k_{-3}}{\overset{k_3}{\rightleftharpoons}} D$$

This describes the autocatalytic production of X and of Y. Again $[F]$ and $[D]$

can be held constant by external constraints when the system is open. In this case only $[X]$ and $[Y]$ are allowed to vary with time and the kinetic equations for this system are

$$\frac{d[X]}{dt} = k_1[F][X] - k_{-1}[X]^2 - k_2[X][Y] + k_{-2}[Y]^2$$

(50)

$$\frac{d[Y]}{dt} = k_2[X][Y] - k_{-2}[Y]^2 - k_3[Y] + k_{-3}[D]$$

We have seen in eqs. (18) and (21) what happens to the extent of reaction when $A \ll RT$. Now, in the opposite limit $A \gg RT$, still from eq. (18) we see that $d\xi/dt \sim v_+$, that is the velocity of reaction approximates the forward velocity of reaction v_+. So the limit $A \gg RT$, which we have said corresponds to conditions far from equilibrium is obtained when the velocities of the reverse reactions, v_-, are negligible that is, in the present case

$$k_{-1}[X]^2, \quad k_{-2}[Y]^2, \quad k_{-3}[D] \ll k_1[F][X], \quad k_2[X][Y], \quad k_3 \qquad (51)$$

If we make this assumption in eq. (50) above, we recover the kinetic equations of the Lotka-Volterra model and so we see that the oscillatory behavior of the Lotka-Volterra model is directly linked to conditions far from equilibrium. We also see that to obtain conditions extremely far from equilibrium, we can simply set the kinetic constants for all reverse reactions to zero. Of course in all these considerations the roles of "forward" and "reverse" can always be interchanged.

We may now wonder how far from equilibrium we should drive the steady state in order to promote oscillatory behavior. Again we can use the simple method of linearizing the kinetic equations in the vicinity of the steady state and inspect the behavior of the deviations from steady state. Let us still use as an example the system given above, and for simplicity let us assume $k_1 = k_2 = k_3 = 1$ and $k_{-1} = k_{-2} = k_{-3} = k$. We find to first order in x and y

$$\frac{dx}{dt} = ([F] - [Y]_{ss} - 2k[X]_{ss})x + (2k[Y]_{ss} - [X]_{ss})y$$

(52)

$$\frac{dy}{dt} = [Y]_{ss}x + ([X]_{ss} - 1 - 2k[Y]_{ss})y$$

where again $[X]_{ss}$ and $[Y]_{ss}$ are the values at steady state which are found by solving the algebraic equations to which the differential equations reduce when $d[X]/dt = d[Y]/dt = 0$. If we evaluate the total affinity A, for the set of coupled reactions we find

$$A = -\ln k^3 \frac{[D]}{[F]} \qquad (53)$$

We recall that for the special steady state represented by the equilibrium state we must have $A = 0$, which gives $k^3[D] = [F]$ as the equilibrium conditions for the reactants F and D. One can easily obtain $[X]_{eq}$ and $[Y]_{eq}$ as $[X]_{eq} = [F]/k$ and $[Y]_{eq} = [F]/k^2$. The linear analysis would show stability under these conditions since deviations from equilibrium are damped. We omit this for brevity, but it can be easily shown by solving the linearized equations for the deviations x and y when $[X]_{ss} = [X]_{eq}$ and $[Y]_{ss} = [Y]_{eq}$.

As we drive F and D from their equilibrium values, A grows and the reaction tends to proceed only in one direction, as

$$\frac{|A|}{RT} \to \infty , \qquad (54)$$

which we have seen to correspond formally to set $k \to 0$. Now we are in a position to analyze the onset of the oscillatory behavior by looking at solutions in the limit $k \to 0$. The first order dependence on k of the steady state concentration is given by

$$[X]_{ss} = 1 + k[F] - k\frac{[D]}{[F]}$$

$$[Y]_{ss} = [F] - k + k[F]^2 \qquad (55)$$

Using these values in the linear equations for the deviations from steady state, x and y, and assuming

$$x = x_0 e^{\lambda t}$$

$$y = y_0 e^{\lambda t} \qquad (56)$$

we get a second degree algebraic equation for λ. The roots of this equation turn out to be real, corresponding to a periodic behavior, for $[D]/[F] \geq \beta$ and complex conjugated, corresponding to oscillatory behavior, for $[D]/[F] < \beta$. The critical value β for the ratio $[D]/[F]$ is found when

$$(1 + [F] + [F]^2 + \beta)^2 k^2 + 4(1 + [D])k - 4[F] = 0 \qquad (57)$$

so that for a given value of k we obtain the critical parameter β or for a given $[D]/[F]$ we get a critical value of k below which oscillatory behavior sets in. Correspondingly we can calculate the critical value of the affinity A beyond which oscillations may start.

So much for studies in time. Let us now briefly explore how structures in space may in principle occur. The simplest example is the autocatalytic reaction

$$F + X \xrightarrow{k} 2X \qquad (58)$$

Let us consider a situation far from equilibrium, and assume that the component X diffuses with diffusion coefficient η and that this diffusion process is much slower than that for component F, which is assumed to be uniformly distributed. For simplicity we restrict our analysis to a single spatial dimension z. Then the equation of continuity for X can be written as

$$\frac{\partial [X]}{\partial t} - \eta \frac{\partial^2 [X]}{\partial z} = k[F][X] \qquad (59)$$

At steady state we have $\partial [X]/\partial t = 0$ and we look for a solution with spatial modulation such as

$$[X] = [X]_0 \sin(\frac{2\pi z}{\lambda} + \phi) \qquad (60)$$

which we obtain if

$$\lambda = 2\pi \sqrt{\frac{\eta}{k[F]}}. \qquad (61)$$

Once again we recall that assuming that the reaction proceeds only in the forward direction is equivalent to assuming that the system is far from equilibrium and we see that under these conditions a spatial pattern can be produced.

A more elaborate example of generation of spatial patterns is seen in the formation of cells of convective flow introduced by a vertical concentration gradient. A polymer solution, a few millimeters thick, is spread over a membrane, swollen with water, which is kept in a horizontal plane. As a small amount of water is released by the membrane, because of the difference in osmotic pressure, the membrane acts as a solvent source, which establishes a vertical gradient of concentration in the polymer solution. Two coupled flows arise in response to the concentration gradient, a diffusive flow and a convective flow. The convective flows are due to the bouyancy forces. These arise from the fact that, due to the release of water by the membrane, the lighter layers find themselves below the denser layers in the gravity field. One can easily see, using the linearized analysis, that as a concentration gradient is imposed beyond a critical volume, an instability occurs and the system displays stationary ascending and descending convective flows, which form an orderly, polygonal, cell-like pattern, Fig. 3.1.

Another feature which is exhibited by living systems is a kind of bistability or multiple stability of different steady states. For instance the sodium and potassium steady state concentrations across a nerve membrane assume high and low steady state values in response to critical values of the electrochemical potential of the membrane. The preceding discussion may have already hinted that while it may prove quite difficult to produce a detailed model of the actual states of a nerve membrane, it may be relatively simple to set up an over-simplified chemical system which displays multiple stability. An instructive example is the following reaction scheme

$$A + 2X \underset{k_{-1}}{\overset{k_1}{\rightleftharpoons}} 3X$$

$$B + X \underset{k_{-2}}{\overset{k_2}{\rightleftharpoons}} C \tag{62}$$

when $[A]$, $[B]$ and $[C]$ are held constant, the kinetic equation for $[X]$ becomes

$$\frac{\mathrm{d}[X]}{\mathrm{d}t} = k_1[A][X]^2 - k_{-1}[X]^3 + k_{-2}[C] - k_2[B][X] \tag{63}$$

Let us set $k_1 = 1$, $k_1[A] = 3$ and write $k_2[B] = \beta$ and $k_{-2}[C] = \gamma$. Then the steady state, that is the situation when $\mathrm{d}[X]/\mathrm{d}t = 0$ is achieved whenever

$$[X]^3 - 3[X]^2 + \beta[X] - \gamma = 0 . \tag{64}$$

This algebraic equation may have either one or three physically acceptable

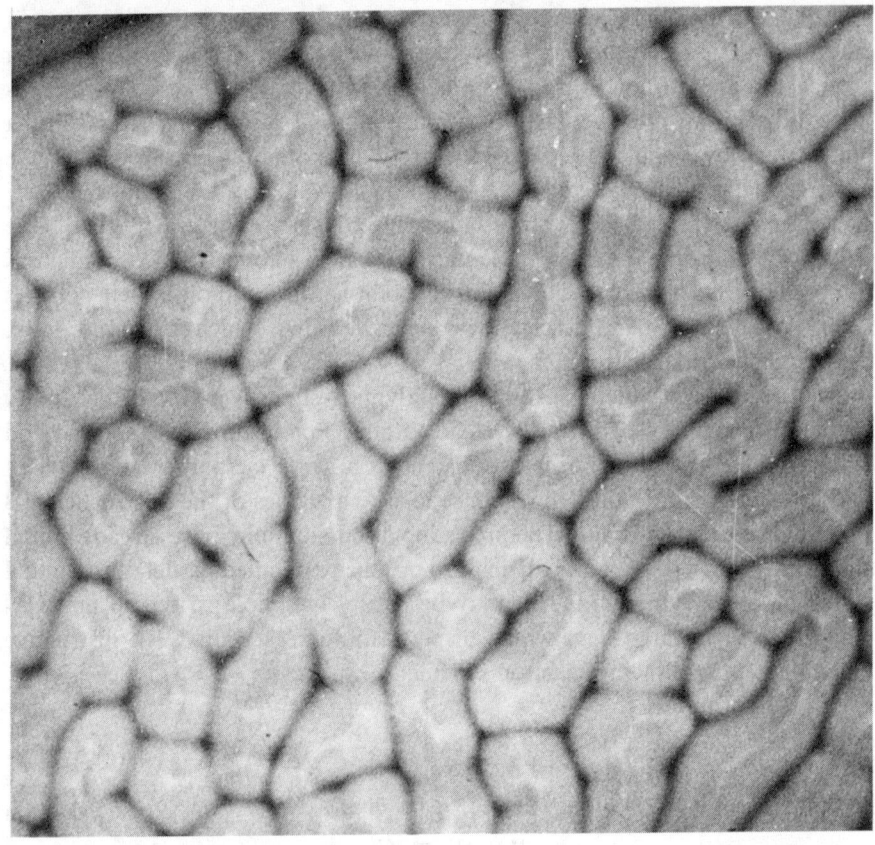

Fig. 3-1 Cell-like pattern of stationary ascending and descending convective flows in 30% solution of polyethylene glycol, 0.2 cm. thick, spread over a dyalisis membrane swollen with water, as seen from above; depressions and elevations of the free surface correspond to descending and ascending flow respectively.

(Reproduced with permission — F. Gambale and G. Gliozzi, J. Phys. Chem. 76, 783 (1972)). Copyright © 1972 American Chemical Society.

solutions for $[X]$ in the real positive domain, depending on the parameters β and γ. Plotting γ versus $[X]$ for various values of β one can see, Fig. 3-2, that for any given γ one has only one solution for $[X]_{ss}$ when $\beta \geq 3$. However, for $\beta < 3$ one obtains three solutions for $[X]_{ss}$. One is unstable, as it can be easily seen by the linearized analysis: any small deviation $[X] - [X]_{ss}$ is found to grow exponentially in time. The other two are stable, as any small deviation will decay exponentially in time to the steady state. Thus the system has two

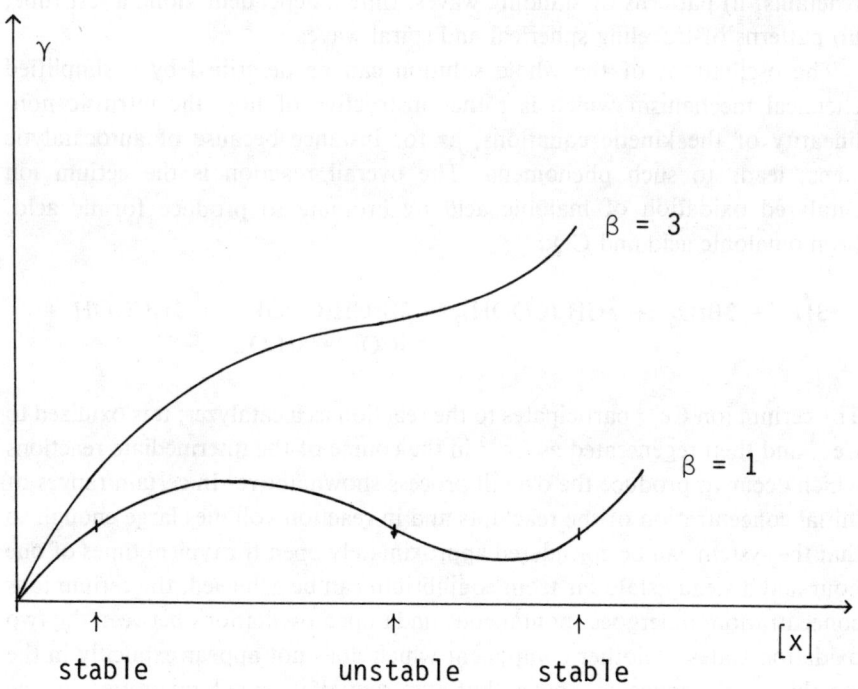

Fig. 3-2 Multiple stability for the steady states provided by the reaction scheme eq. 62, (see text).
(Adapted after F. Schögl, Z. Physik 253, 147 (1972)).

stable states between which it can be switched either by a large fluctuation or by a proper adjustment of β or γ, which depend on the concentrations of reactants $[A]$ and $[C]$.

Real situations, even with relatively simple chemicals, are in practice quite difficult to deal with. Experiments and computer simulations produce a variety of beautiful patterns in space and time. The so called Belousov-Zhabotinskii reaction, the Benard convection and the oscillations in the glycolytic cycle are classic examples of this behavior. We refer the interested reader to the books and reviews listed at the end of this chapter.

Here we call attention to the variety of patterns one can produce experimentally with the Belousov-Zhabotinskii reaction: for the Ce^{+3} ion concentration one observes i) global oscillations in the whole solution; the frequency depends on the temperature of the solution and on the concentrations of the reactants, ii) patterns of standing waves, time independent along a test tube, iii) patterns of traveling spherical and spiral waves.

The oscillations of the whole solution can be described by a simplified chemical mechanism which is rather instructive of how the intrinsic non-linearity of the kinetic equations, as for instance because of autocatalytic steps, leads to such phenomena. The overall reaction is the cerium ion catalyzed oxidation of malonic acid by bromate to produce formic acid, bromomalonic acid and CO_2:

$$3H^+ + 3BrO_3^- + 5CH_2(COOH)_2 \rightarrow 3BrCH(COOH)_2 + 2HCOOH + 4CO_2 + 5H_2O .$$

The cerium ion Ce^{+3} participates to the reaction as a catalyzer: it is oxidized to Ce^{+4} and then regenerated as Ce^{+3} in the course of the intermediate reactions which occur to produce the overall process shown above. In certain ranges of initial concentration of the reactants and in reaction volumes large enough so that the system can be considered approximately open for typical times of one hour and a steady state far from equilibrium can be achieved, the cerium ions concentration undergoes spontaneous undamped oscillations between the two oxidation states. Another component which does not appear explicitly in the stoichiometric equation above, but is essential in small quantities, is the bromide ion Br^-, the concentration of which oscillates concurrently with that of the two cerium ions. The solution turns as a whole from pale yellow, the Ce^{+4} ion, to colourless, with non-sinusoidal oscillation of typical periods of a few minutes. Even the simplest successions of intermediate reactions, proposed to model the overall reaction above, are quite difficult to solve exactly as they involve some twenty interlinked steps. Let us look only at those which give a plausible model of how the oscillations in [Br$^-$] and [Ce^{+3}] occur. Initially we have an appreciable concentration of Br$^-$, so that the dominant reactions are the following, which we call path (*I*)

$$(I) \begin{cases} BrO_3^- + Br^- + 2H^+ \xrightarrow{k_1} HBrO_2 + HOBr \\ \\ HBrO_2 + Br^- + H^+ \xrightarrow{k_2} 2HOBr \end{cases}$$

As $k_1/k_2 \sim 10^{-9}$, the first is the rate limiting step. Also HOBr disappears rapidly because it reacts with malonic acid. Thus we get a quasi-stationary state with

$$[HBrO_2]_{(I)} \sim \frac{k_1}{k_2} [BrO_3^-][H^+]$$

so that the flux in the two reactions is the same. When path (*I*) consumes the Br$^-$, then the dominant reactions become those of path (*II*) according to

$$(II) \begin{cases} BrO_3^- + HBrO_2 + H^+ \xrightarrow{k_3} 2BrO_2 + H_2O \\ BrO_2 + Ce^{+3} + H^+ \xrightarrow{k_4} Ce^{+4} + HBrO_2 \\ 2HBrO_2 \xrightarrow{k_5} BrO_3^- + HOBr + H^+ \end{cases}$$

The first two steps of path (*II*) together represent an autocatalytic step of regeneration of HBrO$_2$. Again the first is the rate limiting step and now the quasi-stationary state has

$$[HBrO_2]_{(II)} \sim \frac{k_3}{k_5} [BrO_3^-][H^+] \ .$$

Looking at the second reaction of path (*I*) and at the first of path (*II*), we see that Br$^-$ and BrO$_3^-$ compete for reacting with HBrO$_2$. Therefore the auto-catalytic regeneration of HBrO$_2$ is not possible as long as

$$k_2[Br^-] > k_3[BrO_3^-]$$

while it becomes possible when this inequality reverses, as Br^- is consumed up in path (I). So there is a critical concentration

$$[Br^-]_c \cong \frac{k_3}{k_2} [BrO_3^-]$$

at which the reaction switches from path (I), because of shortage of Br^-, to path (II). On the other hand the Ce^{+4} ion produced in the course of path (II) regenerates Br^- and Ce^{+3} through the global reaction

$$4Ce^{+4} + BrCH(COOH)_2 + H_2O + HOBr \overset{k_6}{\to} 2Br^- + 4Ce^{+3} + 3CO_2 + 6H^+$$

where some of the bromomalonic acid produced in the other intermediate reactions not shown here is used. We have considered all reactions to proceed only in the forward direction disregarding the extent of the reverse reaction as we assume we are far from equilibrium conditions. Remember also in this context that the concentrations of cerium and bromide are much smaller than those of the main reactants which appear in the overall reaction. Thus we see that this last step induces $[Br^-]$ to rise again, reach above the critical value and switch the reaction back from path (II) to path (I). Concurrently the concentration of Ce^{+4} diminishes to increase that of Ce^{+3}. As we have just said the concentration of Br^- is much smaller than that of malonic acid. So such a switching back and forth between path (I) and path (II) will occur many times, as undamped oscillations, before the concentration of malonic acid and of the other principal reactants will change appreciably. The interested reader can try following this convenient recipe by himself: first dissolve 4.3 gr of malonic acid $CH_2(COOH)_2$ and 0.18 gr of cerium ammonium nitrate $Ce(NO_3)_6$ $(NH_4)_2$ in 150 ml of a 1M solution of sulphuric acid H_2SO_4; keep the solution gently stirred by a suitable apparatus; when the solution again turns clear, after having been yellow for a few minutes, add 1.4 gr of sodium bromate $NaBrO_3$; the solution will turn alternatively clear to yellow, with a period of the order of one minute.

FURTHER READINGS

General

Prigogine, I. (1967) *Introduction to the thermodynamics of irreversible processes,* J. Wiley.

Nicolis, G. and Prigogine, I. (1977) *Self-organization in nonequilibrium systems,* J. Wiley.

Kreuzer, H.J. (1981) *Nonequilibrium thermodynamics and its statistical foundations,* Clarendon Press, Oxford.

Specific

Onsager, L. "Reciprocal relations in irreversible processes I and II", (1931) *Phys. Rev.* **37**, 405 and **38**, 2265.

Prigogine, I., Nicolis, G. and Babloyantz, A. "Thermodynamics of evolution", *Physics Today*, (Nov. 1972) p. 23–28 and (Dec. 1972) 38–44.

Landauer, R. "Inadequacy of entropy and entropy derivatives in characterizing the steady state", (1975) *Phys. Rev.* **A12**, 636.

Prigogine, I. "Time, structure and fluctuations", (1978) *Science* **201**, 777.

Keizer, J. "Nonequilibrium thermodynamics and the stability of states far from equilibrium", (1979) *Acc. Chem. Res.* **12**, 243.

Tyson, J.J. "The Belousov-Zhabotinskii reaction", (1976) *Lecture Notes in Biomathematics* n. 10, Springer Verlag.

Epstein, I.E., Kustin, K., De Kepper, P. and Orban, M. "Oscillating chemical reactions", (March 1983) *Sci. Am.* **248**, 96.

Hess, B., Goldbeter, A. and Lefever, R. "Temporal, spatial and functional order in regulated biochemical and cellular systems", in S. Rice, ed., (1978) *Adv. Chem. Phys.* vol. XXXVIII, J. Wiley.

Jones, B.L., Enns, R.H. and Rangnekar, S.S. "On the theory of selection of coupled macromolecular systems", (1976) *Bull. Math. Biology* **38**, 15.

CHAPTER 4

INTERCONVERSIONS OF CHEMICAL ENERGY: ENZYMATIC CATALYSIS AND ITS CONTROL

> *"We have justifiable reason to suppose that, in living plants and animals, thousands of catalytic processes take place between the tissues and the fluids and result in the formation of the great number of dissimilar chemical compounds, for whose formation out of the common raw material, plant juice or blood, no probable cause could be assigned. The cause will perhaps in the future be discovered in the catalytic power of the organic tissues of which the organs of the living body consist."*
>
> Berzelius, 1836

Living systems are machines. They redirect the flow of energy in order to produce work, achieve chemical synthesis, and create and maintain order. The ultimate source of energy for all living systems is the sun. As will be discussed in the next chapter, the sun is harvested by means of light induced oxidation-reduction reactions which ultimately result in the reduction of CO_2 to form

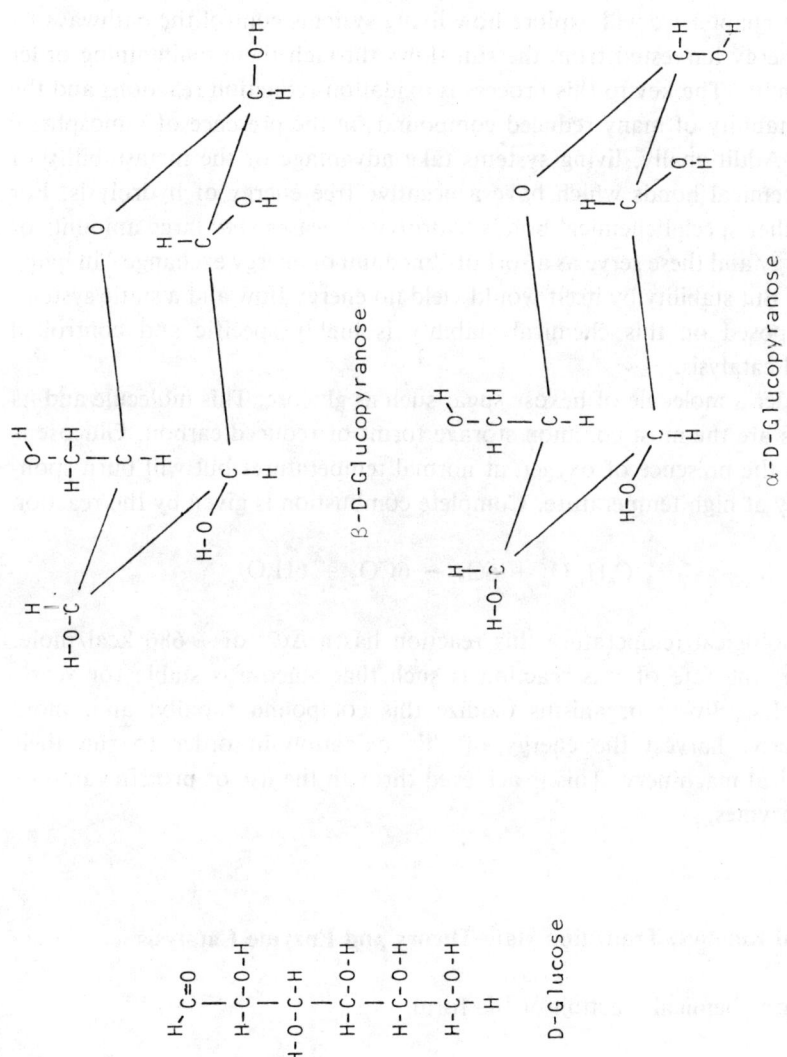

Fig. 4-1 The structures of glucose. In aqueous solution the ring structures predominate.

compounds of reduced carbon such as carbohydrates. Reoxidation of these compounds can liberate this energy and it is this which fuels all living systems, including those which cannot engage directly in photosynthesis.

In this chapter we will explore how living systems control the pathways by which energy harvested from the sun flows through them maintaining order and stability. The key to this process is oxidation-reduction reactions and the kinetic stability of many reduced compounds in the presence of atmospheric oxygen. Additionally, living systems take advantage of the metastability of many chemical bonds which have a negative free energy of hydrolysis. For some rather special chemical bonds hydrolysis releases very large amounts of free energy and these serve as a sort of "medium of energy exchange" in living systems. But stability by itself would yield no energy flow and a static system. Superimposed on this chemical stability is highly specific and controlled chemical catalysis.

Consider a molecule of hexose sugar such as glucose. This molecule and its polymers are the most common storage forms of reduced carbon. Glucose is stable in the presence of oxygen at normal temperatures but will burn spontaneously at high temperature. Complete combustion is given by the reaction

$$C_6H_{12}O_6 + 6O_2 \rightarrow 6CO_2 + 6H_2O$$

At physiological temperature this reaction has a $\Delta G°$ of -686 kcal/mole. However, the rate of this reaction is such that glucose is stable for years. Nevertheless, living organisms oxidize this compound rapidly; and, more importantly, harvest the energy of this oxidation in order to run their biochemical machinery. This is achieved through the use of protein catalysts called enzymes.

Chemical Kinetics, Transition-state Theory and Enzyme Catalysis

Consider a chemical reaction of the form

$$A \underset{k_{-1}}{\overset{k_1}{\rightleftharpoons}} B$$

where A and B are two substances of identical atomic composition and k_1, and k_{-1} are reaction rate constants such that

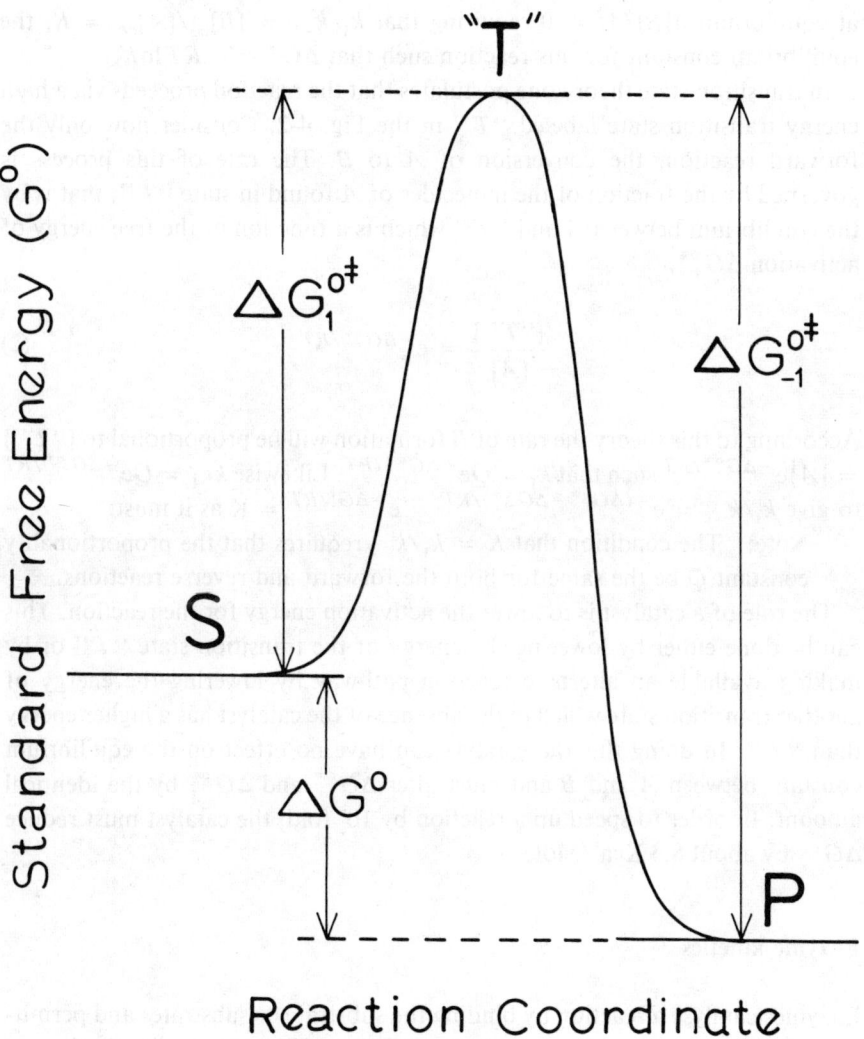

Fig. 4-2 The free energy diagram for transition state theory.

$$\frac{d[B]}{dt} = -\frac{d[A]}{dt} = k_1[A] - k_{-1}[B] \tag{1}$$

at equilibrium $d[B]/dt = 0$ requiring that $k_1/k_{-1} = [B]_{eq}/[A]_{eq} = K$, the equilibrium constant for this reaction such that $\Delta G° = -RT\ln K$.

In transition state theory one postulates that the reaction proceeds via a high energy transition state labeled "T" in the Fig. 4-2. Consider now only the forward reaction, the conversion of A to B. The rate of this process is governed by the fraction of the molecules of A found in state "T", that is by the equilibrium between A and "T" which is a function of the free energy of activation $\Delta G_1^{°\neq}$.

$$\frac{["T"]}{[A]} = e^{-\Delta G_1^{°\neq}/RT} \tag{2}$$

According to this theory the rate of B formation will be proportional to $["T"]$ $= [A]e^{-\Delta G_1^{°\neq}/RT}$ such that $k_1 = Qe^{-\Delta G_1^{°\neq}/RT}$. Likewise $k_{-1} = Qe^{-\Delta G_{-1}^{°\neq}/RT}$ to give $k_1/k_{-1} = e^{-(\Delta G_1^{°\neq} - \Delta G_{-1}^{°\neq})/RT} = e^{-\Delta G°/RT} = K$ as it must.

> Note: The condition that $K = k_1/k_{-1}$ requires that the proportionality constant Q be the same for both the forward and reverse reactions.

The role of a catalyst is to lower the activation energy for the reaction. This can be done either by lowering the energy of the transition state "T" or by making available an alternate reaction pathway by lowering the energy of another transition state which in the absence of the catalyst has a higher energy than "T". In doing this the catalyst can have no effect on the equilibrium constant between A and B and must alter $\Delta G_1^{°\neq}$ and $\Delta G_{-1}^{°\neq}$ by the identical amount. In order to speed up a reaction by 10^6 fold, the catalyst must reduce $\Delta G^{°\neq}$ by about 8.5 Kcal/Mole.

Enzyme kinetics

Enzymes catalyse a reaction by binding the substrate or substrates and permitting by various means the conversion to products. The evidence for the formation of enzyme-substrate complexes is considerable, but is perhaps most clearly implied by the concentration dependencies of the rates of enzyme catalysed reactions.

For classical chemical kinetics, the rate of product formation, dP/dt, depends linearly on the concentration of the reactant. This is also true for higher order reactions so long as they are 1st order with respect to each reactant and the concentration of only a single reactant is varied. A very different situation is found with enzyme catalysed reactions. Here the rate of product formation, at constant enzyme concentration, varies with substrate concentration by a saturable function which follows the equation of a rectangular hyperbola, i.e.

$$\frac{dP}{dt} = v = \frac{V_{max}[S]}{[S] + K_m} \qquad (3)$$

where V_{max} is the maximum rate of product formation achieved at high substrate concentration, $[S]$, and K_m is a constant with the same units as $[S]$. It was Michaelis and Menton who put forth the first simple formulation of enzyme catalysis which accounted for this behavior. For a single substrate reaction they postulated the formation of an enzyme-substrate complex as an intermediate in the formation of product.

$$E + S \underset{k_{-1}}{\overset{k_1}{\rightleftharpoons}} ES \underset{k_{-2}}{\overset{k_2}{\rightleftharpoons}} E + P$$

This system can be analysed trivially if one makes two simplifying assumptions; first, that $[S]$, the substrate concentration, does not vary significantly during this rate measurement and second, that the measurement is made at the beginning of the reaction so that $[P]$, the concentration of product is small and the effects of k_{-2} can be ignored.

Such a system, upon mixing E with S will relax to a steady state in which $d[ES]/dt = 0$ and one can therefore write that

$$[E][S]k_1 = [ES](k_{-1} + k_2) \qquad (4)$$

or

$$\frac{[E][S]}{[ES]} = \frac{k_{-1} + k_2}{k_1} = K_m . \qquad (5)$$

Although this system is not at equilibrium, but rather at steady state, an important distinction, this is the equation of a simple binding isotherm. If the total enzyme concentration, E_T, is given by

$$E_T = [E] + [ES] \tag{6}$$

then

$$\frac{[ES]}{E_T} = \frac{[S]}{K_m + [S]} \tag{7}$$

and noting that

$$v = \frac{d[P]}{dt} = k_2[ES] \quad \text{and} \quad V_{max} = k_2 E_T \tag{8}$$

one can write

$$v = \frac{V_{max}[S]}{K_m + [S]} = V_{max} \frac{[S]/K_m}{1 + ([S]/K_m)} = V_{max} \frac{1}{(K_m/[S]) + 1} \tag{9}$$

There are a number of things to be noted about this equation which are of significance to living systems. The value of V_{max} can be varied by varying E_T, the amount of enzyme present, but K_m is a property of the enzyme. The dependence of v on $[S]$ varies with the ratio $[S]/K_m$. For $[S]/K_m \ll 1$ the relationship is approximately linear. For $[S]/K_m \gg 1$, $v = V_{max}$ and the dependence on $[S]$ is lost. Therefore, if it is advantageous for the rate of product formation to be modulated by the substrate concentration, then an enzyme should have a K_m value which is higher than the substrate concentrations it is likely to encounter.

Enzyme catalysis of thermodynamically favorable but kinetically inhibited reactions is a powerful tool by which living systems utilize chemical energy for chemical synthesis, mechanical work or the generation of order. There are many pathways which one can envision for the oxidation of glucose to CO_2 and water. If one simply catalysed that conversion, by whatever pathway, little would be accomplished except to generate heat. However, that is not what is done. Instead, this conversion is achieved by a complex, multistep pathway in which certain steps, which are thermodynamically very favorable, are coupled to thermodynamically unfavorable reactions to produce compounds in which chemical energy can be stored but from which it can be quickly extracted to promote chemical synthesis or mechanical work. Sometimes this coupling process is quite direct, for example the use of phosphorolysis instead of hydrolysis to cleave a bond. In others it is much more indirect, being mediated by the generation of electro-chemical gradients which are then used to promote the second reaction. Examples of both will be discussed.

Glycolysis

Let us now examine some of the metabolic pathways by which living organisms utilize glucose. For the moment we shall focus on the glycolytic pathway by which glucose is converted anaerobically to lactate.

$$
\begin{array}{ccc}
\text{CHO} & & \\
| & & \text{O} \\
\text{HCOH} & & || \\
| & & \text{COH} \\
\text{HOCH} & \longrightarrow & | \\
| & & 2\,\text{HCOH} \\
\text{HCOH} & & | \\
| & & \text{CH}_3 \\
\text{HCOH} & & \\
| & & \\
\text{CH}_2\text{OH} & &
\end{array}
$$

This conversion involves no net oxidation, merely the cleavage of the molecule into two parts and some rearrangements. The standard free energy for the reaction as written is -45 kcal/mole. The overall reaction carried out by most living organisms is rather different from this, in that the conversion of glucose to lactic acid is coupled to the formation of two adenosine triphosphate (ATP) molecules from adenosine diphosphate (ADP) and inorganic phosphate (Pi).

$$\text{Glucose} + 2\text{ADP} + 2\text{Pi} \rightarrow 2 \text{ Lactate} + 2\text{ATP} + 2\text{H}_2\text{O}$$

The ATP molecule whose structure is shown in Fig. 4-3 is of central importance throughout biological energy metabolism. This is due to the fact that the phosphodiester linkages of this compound possess a large negative free energy of hydrolysis

$$\text{ATP} + \text{H}_2\text{O} \rightarrow \text{ADP} + \text{Pi} \qquad \Delta G^\circ = -7.3 \text{ kcal/mole}$$

Therefore in glycolysis, 32% of the free energy of the process (at standard state) is harvested as ATP.

The glycolytic pathway is illustrated in Fig. 4-4. It begins with the investment of two ATP molecules, first to phosphorylate glucose and then,

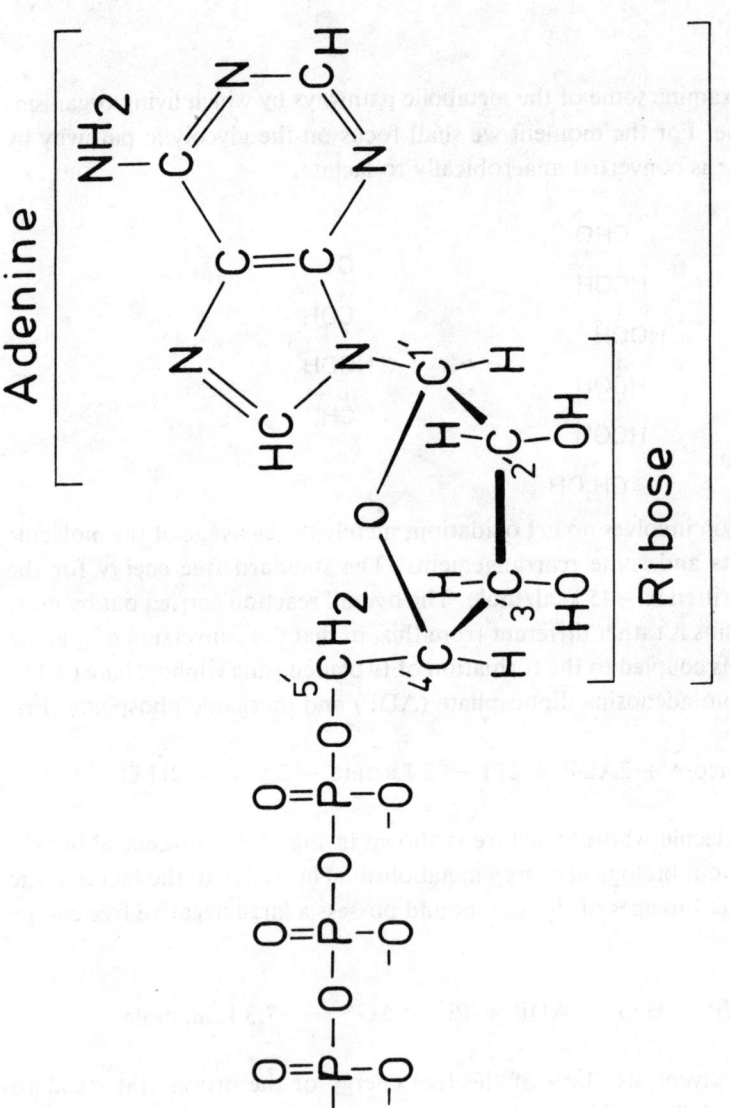

Fig. 4-3 The structure of adenosine triphosphate, ATP. Adenosine diphosphate, ADP, and adenosine monophosphate, AMP, contain 2 and 1 phosphate groups respectively.

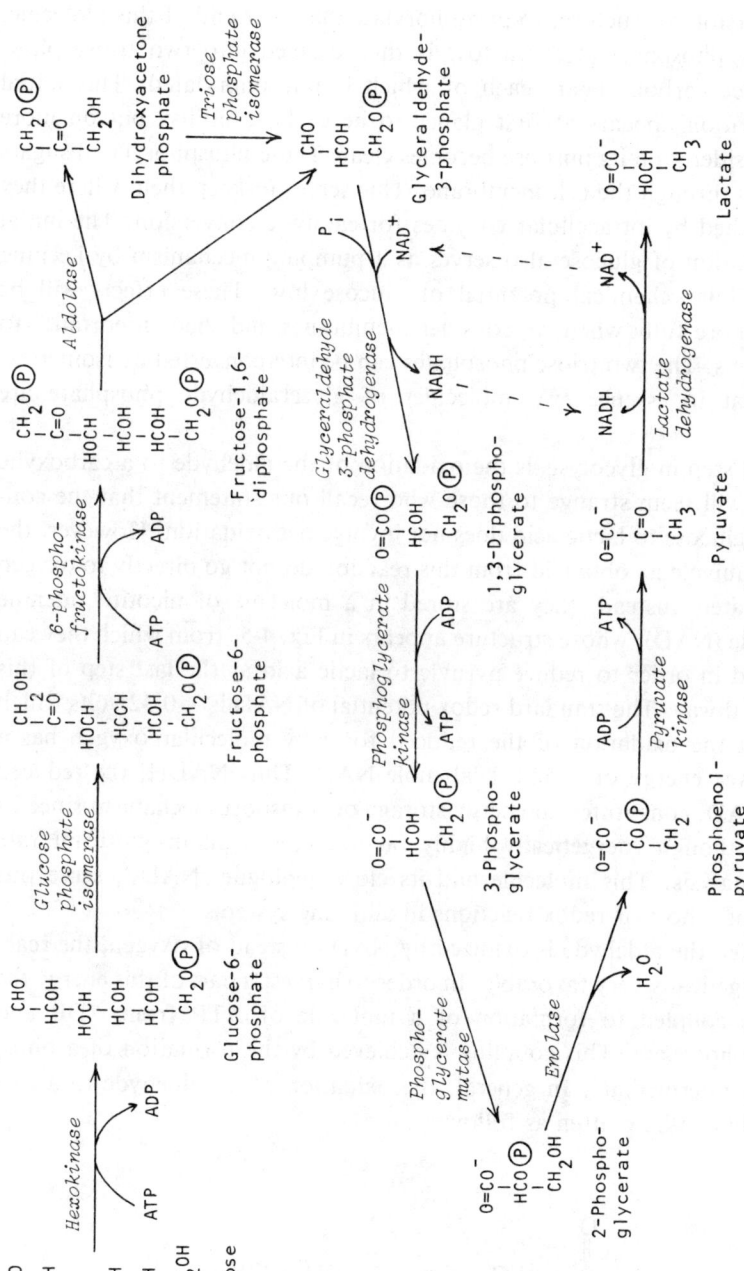

Fig. 4-4 An outline of anaerobic glycolysis, the pathway for the conversion of a molecule of glucose to two molecules of lactate with the concomitant formation of two molecules of ATP from ADP and inorganic phosphate.

after conversion to fructose, to phosphorylate the other end of this molecule. The doubly phosphorylated fructose is then cleaved into two triose phosphates, three carbon sugars each of which is phosphorylated. This initial phosphorylation appears at first glance to be of little utility, but on more careful consideration its purpose becomes clear. These phosphorylated sugars cannot pass through the cell membrane. This serves to keep them where they can be reached by intracellular enzymes for catalytic conversion. The initial phosphorylation of glucose also serves as a pumping mechanism by keeping the intracellular chemical potential of glucose low. These effects will be examined more fully when we consider membranes and their importance to living systems. The two triose phosphates can be interconverted by isomerization so that in essence two molecules of glyceraldehyde phosphate are produced.

The next step in glycolysis is the oxidation of the aldehyde to a carboxylic acid. This will seem strange to those who recall our statement that the conversion of glucose to lactic acid does not involve net oxidation. However, the reducing equivalents obtained from this reaction do not go directly to oxygen to form water. Instead, they are stored in a molecule of nicotine adenine dinucleotide (NAD), whose structure appears in Fig. 4-5, from which they can be obtained in order to reduce pyruvic to lactic acid as the last step of this reaction pathway. The standard redox potential of NAD is -0.32 volts which means that the oxidation of the reduced form by molecular oxygen has a standard free energy of -52.6 kcal/mole NAD. Thus NADH, the reduced form of NAD, constitutes an energy storage or transport mechanism since its oxidation, though energetically highly favorable, has an insignificant rate without catalysis. This molecule and its close analogue, NADP, serve this function for a host of redox reactions in all living systems.

Even when the aldehyde is oxidized by NAD^+ instead of oxygen, the reaction is energetically very favorable. In order to harvest a part of this energy the reaction is coupled to formation of a molecule of ATP from ADP and inorganic phosphate. This coupling is achieved by the formation of a phosphorylated intermediate. In general, the oxidation of an aldehyde to a carboxylic acid can be written as follows

$$
\begin{array}{c}
\overset{\textstyle O}{\overset{\|}{H-C}} + H_2O \longrightarrow \overset{\textstyle O}{\overset{\|}{C}}-OH + 2H \cdot \\
\underset{\textstyle R}{|} \qquad\qquad \underset{\textstyle R}{|}
\end{array}
$$

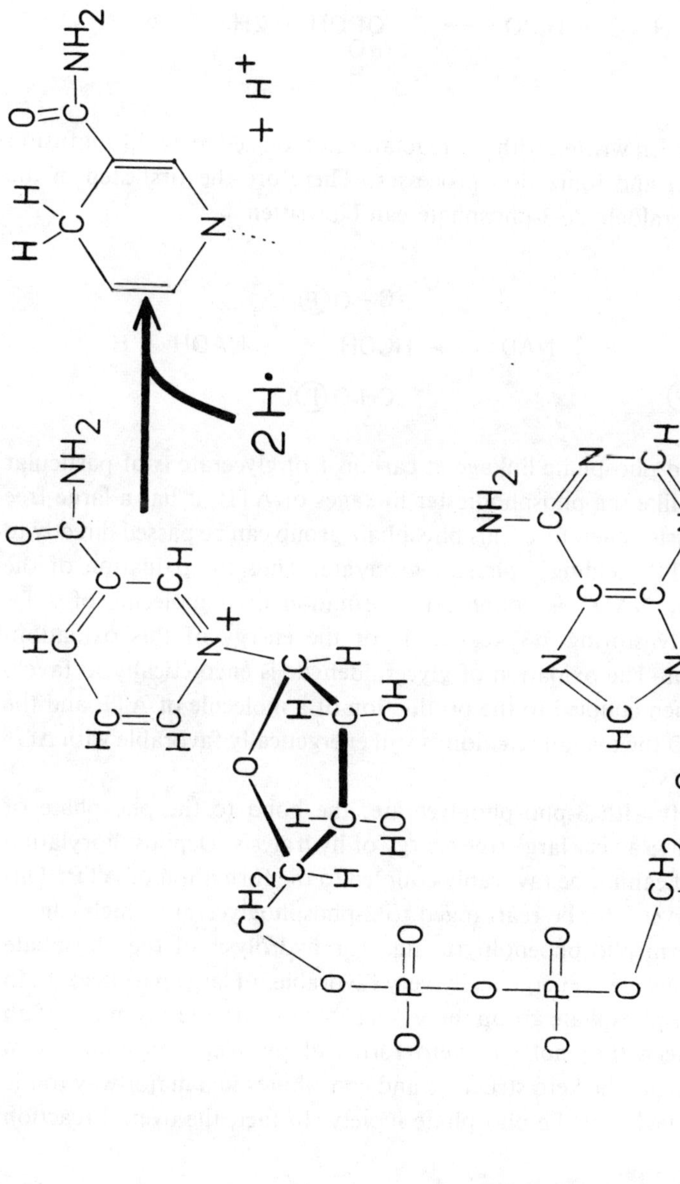

Fig. 4-5 The structure of nicotinamide adenine dinucleotide, NAD. The structural difference between the oxidized, NAD+, and reduced, NADH, forms is shown.

where the hydrogen radicals reduce NAD or oxygen or some other oxidizing agent. Instead of this reaction, biological systems use a phosphate ion in place of a water molecule for the oxidation of glyceraldehyde-3-phosphate to give the reaction

$$\underset{\substack{\displaystyle | \\ R}}{\overset{\displaystyle \overset{\textstyle O}{\|}}{H-C}} + H_3PO_4 \longrightarrow \underset{\substack{\displaystyle | \quad | \\ R \quad H}}{\overset{\displaystyle \overset{\textstyle O}{\|} \;\; \overset{\textstyle O}{\|}}{C-OPOH}} + 2H \cdot$$

The reaction has been written with the reactants not ionized to avoid confusion between oxidation and ionization processes. Therefore the first step in the oxidation of glyceraldehyde-3-phosphate can be written

$$\begin{array}{c} \text{HC} \overset{\displaystyle \nearrow O}{} \\ | \\ \text{HCOH} \\ | \\ \text{CH}_2\text{O}\,\textcircled{P} \end{array} + P_i + NAD^+ \longrightarrow \begin{array}{c} O \overset{\displaystyle \searrow}{} \text{C}-\text{O}\,\textcircled{P} \\ | \\ \text{HCOH} \\ | \\ \text{CH}_2\text{O}\,\textcircled{P} \end{array} + NADH + H^+$$

The newly formed phosphate linkage at carbon 1 of glycerate is of particular interest because, like the phosphodiester linkages of ATP, it has a large free energy of hydrolysis. Therefore, this phosphate group can be passed directly to ADP to form ATP yielding 3-phosphoglycerate. Thus the oxidation of the glyceraldehyde by NAD^+ is coupled to formation of a molecule of ATP thereby chemically storing 7.3 kcal/mole of the energy of this oxidation-reduction reaction. The oxidation of glyceraldehyde is energetically so favorable that even when coupled to the production of a molecule of ATP and the reduction of NAD the overall reaction is still energetically favorable with $\Delta G° = -3$ kcal/mole.

One is now left with 3-phosphoglycerate, the bond to the phosphate of which does not have a very large free energy of hydrolysis. Dephosphorylation of this compound cannot be favorably coupled to the formation of ATP. This compound, however, can be rearranged to 2-phosphoglycerate which can be dehydrated to form phosphoenolpyruvate. The hydrolysis of the phosphate linkage in this molecule is energetically very favorable, in large part because in the absence of the phosphate group there is a spontaneous establishment of an equilibrium between the enol and keto forms of pyruvate, an equilibrium which strongly favors the keto structure and contributes in a major way to the free energy of removal of the phosphate moiety. In fact, the overall reaction

$$
\begin{array}{ccc}
\underset{C}{\overset{O}{\diagup}}-O^- & \underset{C}{\overset{O}{\diagup}}-O^- & \underset{C}{\overset{O}{\diagup}}-O^- \\
| & | & | \\
C-O-\textcircled{P} \;+\; H_2O \;\longrightarrow & C-OH \;\Longrightarrow & C=O \\
\| & \| & | \\
CH_2 & CH_2 & CH_3 \\
& + & \\
& P_i &
\end{array}
$$

for the hydrolysis of PEP to form pyruvate has a standard free energy, ΔG°, of -14.8 kcal/mole, more than adequate to be coupled to the formation of ATP.

In the absence of oxygen, the reducing equivalents on NADH must be passed to some other molecule in order to reform NAD^+ thus permitting subsequent oxidation of glyceraldehyde-3-phosphate. In mammalian tissues this is accomplished by reducing pyruvate to lactate. There are alternative pathways such as the decarboxylation of pyruvate to form acetaldehyde and the reduction of this compound to form ethanol, a pathway used by microorganisms in the formation of wine, etc., but in all cases NAD^+ must be regenerated. This is because the total amount of NAD in the system is limited.

If oxygen is present and the organism is capable of oxidative metabolism, then the reducing equivalents from NADH are passed to oxygen through an electron transport chain with the concomitant formation of ATP by a process termed oxidative phosphorylation. We will have much more to say about this process later in this book. However, at present we want to consider the glycolytic pathway as a model for the control of metabolic processes.

The reasons that mechanisms for the regulation of a metabolic pathway, such as glycolysis, are necessary are obvious if one recalls the dynamic nature of living organisms and the role this pathway is designed (or evolved) to play. Glycolysis produces ATP under anaerobic conditions. Under aerobic conditions it is a major source of pyruvate which can be oxidized to CO_2 and water by another metabolic cycle, the Kreb's or citric acid cycle, producing large amounts of ATP both directly and from the subsequent oxidation of the NADH thus formed. Therefore, the change from anaerobic to aerobic metabolism greatly alters this need for glycolytic activity since in the latter situations ATP is available from another source. To make matters even more complex, there are sources of metabolites for the citric acid cycle other than sugars. Proteins, or rather their constituent amino acids can be converted to pyruvate and also to acetyl coenzyme A, the next intermediate from pyruvate leading to the citric acid cycle. Fatty acids are another source of acetyl coenzyme A. Therefore, the need for glycolysis under aerobic conditions will vary with the availability of alternative metabolites.

Modulation of enzyme activity

Inhibitors which bind reversibly

Frequently there are substances which can bind to an enzyme and interfere with its catalytic activity. A common type of inhibitor is one which competes with the substrate for its binding site on the enzyme.

$$E + S \underset{k_{-1}}{\overset{k_1}{\rightleftharpoons}} ES \overset{k_2}{\rightarrow} E + P$$

$$+$$

$$I$$

$$\Updownarrow K_I$$

$$EI$$

If K_I is the dissociation constant for the Enzyme-Inhibitor complex then one can show that

$$v = \frac{V_{max}[S]}{[S] + K_m(1 + \dfrac{[I]}{K_I})} \tag{10}$$

This is termed a competitive inhibitor and has the effect of altering the apparent value of K_m, $K_m^{App} = K_m(1 + [I]/K_I)$, without affecting V_{max}.

Alternatively, one can have an inhibitor which does not affect substrate binding but which blocks the catalytic process. Such an inhibitor reacts as follows:

$$I + E + S \underset{k_{-1}}{\overset{k_1}{\rightleftharpoons}} I + ES \overset{k_2}{\rightarrow} E + P$$

$$\Updownarrow K_I \qquad\qquad\qquad \Updownarrow K_I'$$

$$EI + S \qquad \rightleftharpoons \qquad EIS$$

If the inhibitor truly has no effect on substrate binding, then $K_I = K_I'$ and it can be shown that

$$v = \frac{[S]V_{max}}{(K_m + [S])} \frac{1}{(1 + \frac{[I]}{K_I})} \tag{11}$$

This is termed a non-competitive inhibitor and has the effect of altering the apparent value of V_{max}, $V_{max}^{App} = V_{max}/(1 + [I]/K_I)$, without affecting K_m. Finally, one can have situations in which $K_I \neq K_I'$. In such a case the inhibitor affects substrate binding without completely blocking it and the apparent values of K_m and V_{max} are both altered.

In all cases the inhibitor reacts with the enzyme by a simple binding isotherm, i.e., by a hyperbolic function. Therefore the change in the inhibitor concentration required to produce a sizable change in v is large.

These inhibitors which we have described act by binding at the active site of the enzyme. They are therefore limited to those compounds which either resemble structurally the substrate (competitive inhibitors) or which interact with residues required for catalytic activity. This severely limits the number of different compounds that can inhibit a particular enzyme by such a mechanism. In the case of the metabolic pathway we have been discussing, it is clear that it would be advantageous to have the pathway controlled by, for example, the concentration of ATP or of pyruvate or of citrate, the first component of the Krebs cycle. Furthermore, we will in general want to inhibit not the last enzyme in a metabolic pathway but the first in order to avoid the accumulation of unnecessary metabolic intermediates. The likelihood that by chance the products of a pathway will act effectively as competitive or classical non-competitive inhibitors of an enzyme catalysing an early step in the pathway is small. To achieve this type of control another class of inhibitors is required, the allosteric inhibitors.

Allosteric inhibition

So far we have discussed the inhibition of enzymes by compounds which bind at the active site, thereby blocking either substrate binding, catalysis, or both. An alternative to this mechanism is the binding of a molecule to the enzyme resulting in a conformational change in the enzyme, thereby altering its substrate binding affinity or catalytic activity. This situation is pictured diagrammatically in the Fig. 4-6. Here we picture the enzyme as having two possible

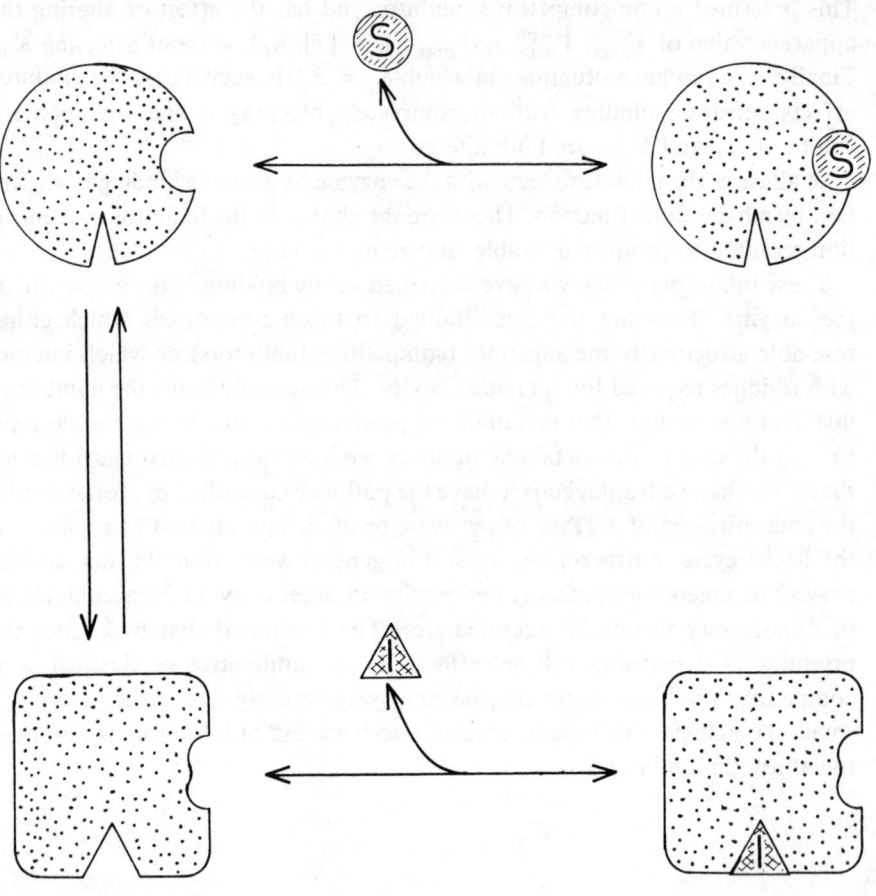

Fig. 4-6 A possible mechanism for allosteric inhibition. The inhibitor, *I,* binds to a conformational state of the protein to which the substrate, *S,* binds poorly.

structures or conformations, one which binds the substrate well but the inhibitor poorly, the second which binds the inhibitor well but the substrate poorly. In drawing such a diagram one necessarily postulates a mechanism for the phenomenon which we are discussing. This should be realized and the diagram taken as a possible explanation, not as a general formulation of mechanism. The postulations of a conformational change as a part of the mechanism is almost certainly correct, but in the most general formulation the binding of inhibitor or substrate might be postulated to alter the distribution function describing the conformational equilibria of the protein.

The thermodynamic formulation of this phenomenon requires no postulate of mechanism, nor does it offer one, but it demonstrates some interesting properties of such systems. Consider a molecule with two binding sites, one for a molecule, I, and the second for a molecule, S, where S and I are dissimilar. Both S and I can bind simultaneously. The binding reactions possible in this system are as follows.

$$E + S + I \underset{K_S}{\rightleftharpoons} ES + I$$

$$K_I \updownarrow \qquad\qquad \updownarrow K_{IS}$$

$$EI + S \underset{K_{SI}}{\rightleftharpoons} ESI$$

The K's are the equilibrium constants for the individual reactions and throughout this treatment will be taken to be the constants for the dissociation reaction, i.e. $K_S = [E][S]/[ES]$ at equilibrium. Since the equilibrium constant, or free energy difference, between two states must be independent of path, it is clear that

$$K_S \cdot K_{IS} = K_I \cdot K_{SI} \tag{11}$$

or

$$\frac{K_{SI}}{K_S} = \frac{K_{IS}}{K_I} \tag{12}$$

This demonstrates the reciprocal nature of such a system. The ratio of the dissociation constants for S in the presence and absence of I must equal the ratio of the dissociation constants for I in the presence and absence of S. If we consider instead the free energies of these reactions,

$$\Delta G = \Delta G^\circ + RT \ln \frac{\Pi_i C_i^{n_i}}{\Pi_j C_j^{n_j}} \qquad (13)$$

where the i components are the products of the reaction and j refer to the reactant or reactants, it is trivial to show that

$$\Delta G_{SI} - \Delta G_S = \Delta G_{IS} - \Delta G_I \qquad (14)$$

The effect of I on the free energy of the interaction of S with E must equal the effect of S on the free energy of the interaction of I with E. Again the system is completely symmetrical.

The analysis given above considers only the effect of saturating the protein with an effector ligand on the binding of the second ligand. Let us now consider the situation when the effector ligand is not present at saturating concentrations. Under these circumstances all possible components of the reaction scheme may exist in solution. Assuming the system to be at equilibrium, one can write the equation for the apparent dissociation constant of S at a given concentration of I as

$$\bar{K}_S = \frac{[S]([E] + [EI])}{[ES] + [ESI]} \qquad (15)$$

but $[EI] = [E][I]/K_I$ and $[EIS] = [ES][I]/K_{IS}$. Therefore

$$\bar{K}_S = \frac{[S][E](1 + \dfrac{[I]}{K_I})}{[ES](1 + \dfrac{[I]}{K_{IS}})} = K_S \frac{(1 + \dfrac{[I]}{K_I})}{(1 + \dfrac{[I]}{K_{IS}})} \qquad (16)$$

Note that when $K_I = K_{IS}$, $\bar{K}_S = K_S$ as it does when $[I] = 0$.

For $[I] \gg K_I, K_{IS}$

$$\bar{K}_S = K_S \frac{K_{IS}}{K_I} = K_{SI} \qquad (17)$$

Taking the logarithm of the equation for K_S we obtain

$$\log \overline{K}_S = \log K_S + \log (1 + [I]/K_I) - \log (1 + [I]/K_{IS}) \qquad (18)$$

Differentiating with respect to $\log [I]$ yields

$$\frac{d \log \overline{K}_S}{d \log [I]} = \frac{[I]}{K_I + [I]} - \frac{[I]}{K_{IS} + [I]} \qquad (19)$$

The first term is equal to the fractional saturation of the enzyme with I at the concentration given in the absence of S, Y_I^O. The second is the fractional saturation of the protein with I when it is saturated with S, Y_I^S. The difference, $Y_I^O - Y_I^S$ is equal to the number of moles of I which dissociate from a mole of E in response to the binding of S, ΔY_I. The absolute value of ΔY_I is at a maximum when $[I] = \sqrt{K_I K_{IS}}$ and approaches unity for $K_I \gg K_{IS}$ or $K_I \ll K_{IS}$. $\log K_S$ varies with $\log [I]$ only when the dissociation constant for I is affected by the binding of S. $d \log K_S/d \log [I]$ is non zero only when the system is not saturated with I which means that the amount of I bound will change when the dissociation constant changes as a result of binding S.

It is perhaps worth noting that

$$\frac{d \log \overline{K}_S}{d \log [I]} = -\frac{d\Delta \overline{G}_S^\circ}{d\mu_I} \qquad (20)$$

where $\Delta G^\circ{}_S$ is the apparent standard free energy of the dissociation of S from E and μ_I is the chemical potential of I. In Fig. 4-7 $\log \overline{K}_S$ and $d \log \overline{K}_S/d \log [I]$ are plotted as a function of $\log [I]$ for a system for which $K_S = 10^{-4}$, $K_{SI} = 10^{-1}$, and $K_I = 10^{-5}$.

These are symmetrical functions, $\log \overline{K}_S$ being a symmetrical function of $\log [I]$ around the point $\log \overline{K}_S = (\log K_S K_{SI})/2$ and $\log [I] = \log (K_I K_{IS})^{1/2}$ while $d \log \overline{K}_S/d \log [I]$ is symmetrical about the line defined by the same value of $\log [I]$. Throughout this derivation I and S can be interchanged in order to give a similar function for $\overline{K}_I$ and $d \log \overline{K}_I/d \log [S]$. The above analysis can only deal with variations in binding affinities. It is also quite possible for a change in the conformation of an enzyme to alter its catalytic rate. However, this would be an effect on V_{max} not K_S and would not affect our thermodynamic analysis. However, it could increase the metabolic control exercised by the inhibitor. In addition, it is clear that the binding of I can as easily increase the affinity of the protein for S as decrease it. In the former case I would be an allosteric activator rather than an inhibitor. In this case these

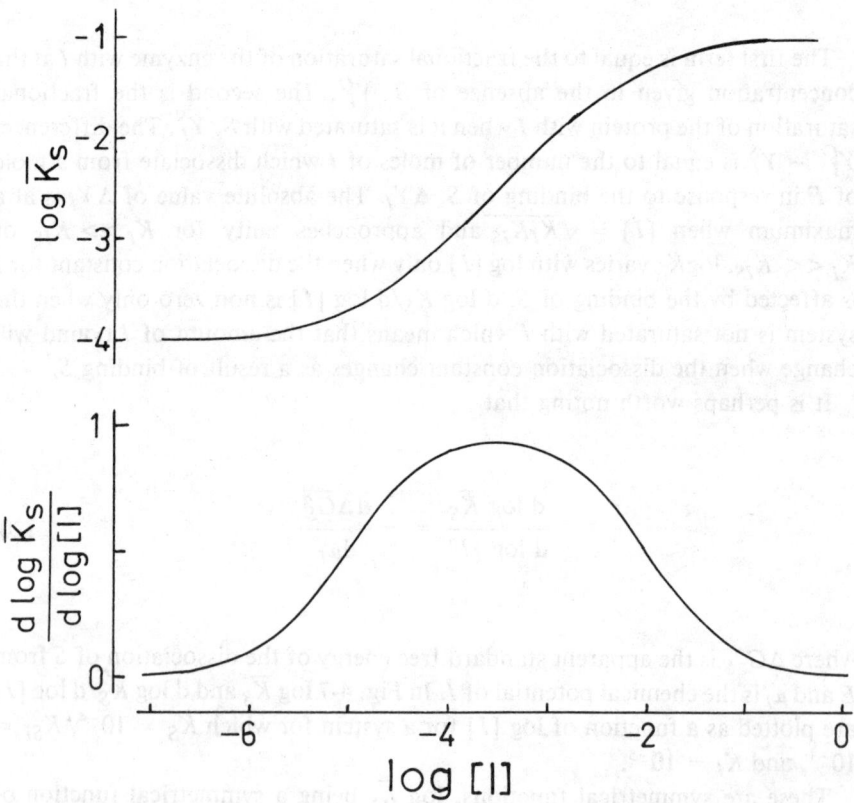

Fig. 4-7 The dependence of $\log \bar{K}_s$ and d $\log \bar{K}_s/$d $\log [I]$ on $\log [I]$ when $K_s = 10^{-4} M$, $K_{SI} = 10^{-1} M$, and $K_I = 10^{-5} M$.

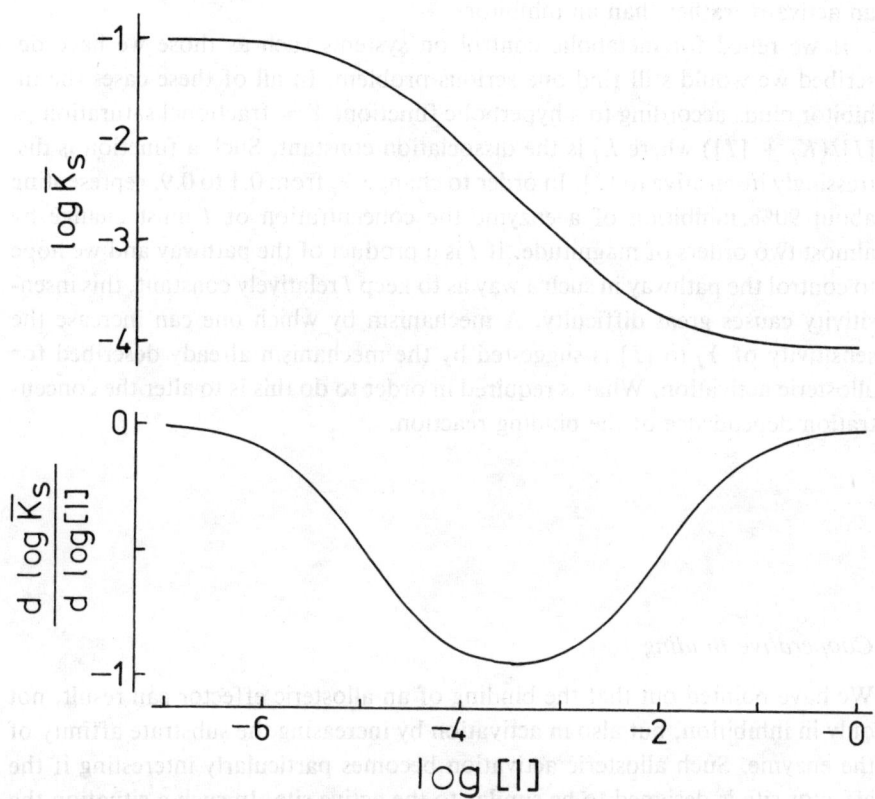

Fig. 4-8 The dependence of $\log \bar{K}_S$ and $d \log \bar{K}_S / d \log [I]$ on $\log [I]$ when $K_S = 10^{-1} M$, $K_{SI} = 10^{-4} M$, and $K_I = 10^{-2} M$.

dependencies of $\log \bar{K}_S$ and $d \log \bar{K}_S / d \log I$ on $\log I$ would appear as in Fig. 4-8.

What does one gain from such considerations? First we can now see a mechanism by which an enzyme can be designed to be inhibited or have its activity modulated by virtually any substance. The binding site for the allosteric effector need not conform to the needs to bind the substrate or any molecule other than the effector. We have also seen how an effector molecule can act as an activator rather than an inhibitor.

If we relied for metabolic control on systems such as those we have described we would still find one serious problem. In all of these cases the inhibitor binds according to a hyperbolic function. Y = fractional saturation = $[I]/(K_I + [I])$ where K_I is the dissociation constant, Such a function is distressingly insensitive to $[I]$. In order to change Y_I from 0.1 to 0.9, representing about 90% inhibition of a enzyme the concentration of I must change by almost two orders of magnitude. If I is a product of the pathway and we hope to control the pathway in such a way as to keep I relatively constant, this insensitivity causes great difficulty. A mechanism by which one can increase the sensitivity of Y_I to $[I]$ is suggested by the mechanism already described for allosteric activation. What is required in order to do this is to alter the concentration dependence of the binding reaction.

Cooperative Binding

We have pointed out that the binding of an allosteric effector can result, not only in inhibition, but also in activation by increasing the substrate affinity of the enzyme. Such allosteric activation becomes particularly interesting if the effector site is designed to be similar to the active site. In such a situation the binding of one molecule can act to facilitate the binding of a second, identical molecule resulting in cooperative binding.

Generally, if nature needs two sites on a protein that bind the same structure, the second site is not developed *de novo*, but is instead derived from the first. The simplest way to achieve multiple, identical binding sites on a protein is to allow the monomeric protein to evolve in such a way that the monomers bind together to form a multisubunit protein. Thus, as a general rule, proteins displaying cooperative binding are composed of multiple subunits. In such proteins the conformational effects of the binding of a molecule must be transmitted from one subunit to another through the interface of their mutual contact. This may at first seem unlikely, but one should recall several facts before dismissing this possibility. First, the structures of even proteins with but a single subunit are determined primarily by the same noncovalent interactions which link the subunits of the multisubunit protein. In addition, the energetics of protein folding is greatly affected by interactions at the surface of the protein. Secondly, the contacts between such subunits are not the point contacts one would envision between two spheres, but instead involve large regions of the surfaces of each subunit.

Let us consider the simplest situation which permits cooperative ligand binding, a protein containing two identical binding sites. In Fig. 4-9 we see a pictorial representation of a possible mechanism for cooperative binding. Two subunits interact to produce a protein with a two-fold axis of rotational symmetry. In the absence of ligand, S, both subunits have identical structures which can bind S only with difficulty. The binding of S to one subunit causes a structural change which is transmitted to the second subunit, in this case maintaining the structural symmetry. The second subunit now has a binding site which readily binds S. Thus the affinity for binding the first S is less than that for the second binding event. K_1 and K_2 being dissociation constants we can write $K_1 > K_2$.

It should be noted that even if the two binding sites were identical and no structural changes occurred on ligand binding K_1 and K_2 would not be equal. In fact under these conditions

$$K_2 = 4K_1 = 2K_{int} \qquad (21)$$

where K_{int} is the intrinsic dissociation constant of any particular single binding site. This is due to statistical factors which result from our not distinguishing between the two binding sites.

The equation for the fractional saturation, Y, as a function of S is

$$Y = \frac{[PS] + 2[PS_2]}{2([P] + [PS] + [PS_2])} = \frac{\dfrac{[S]}{K_1} + 2\dfrac{[S]^2}{K_1 K_2}}{2\left(1 + \dfrac{[S]}{K_1} + \dfrac{[S]^2}{K_1 K_2}\right)} \qquad (22)$$

If $K_1 = K_2/4$, i.e. there is no cooperativity, this equation simplifies to

$$Y = \frac{[S]/K_{\text{int}}}{1 + [S]/K_{\text{int}}} \tag{23}$$

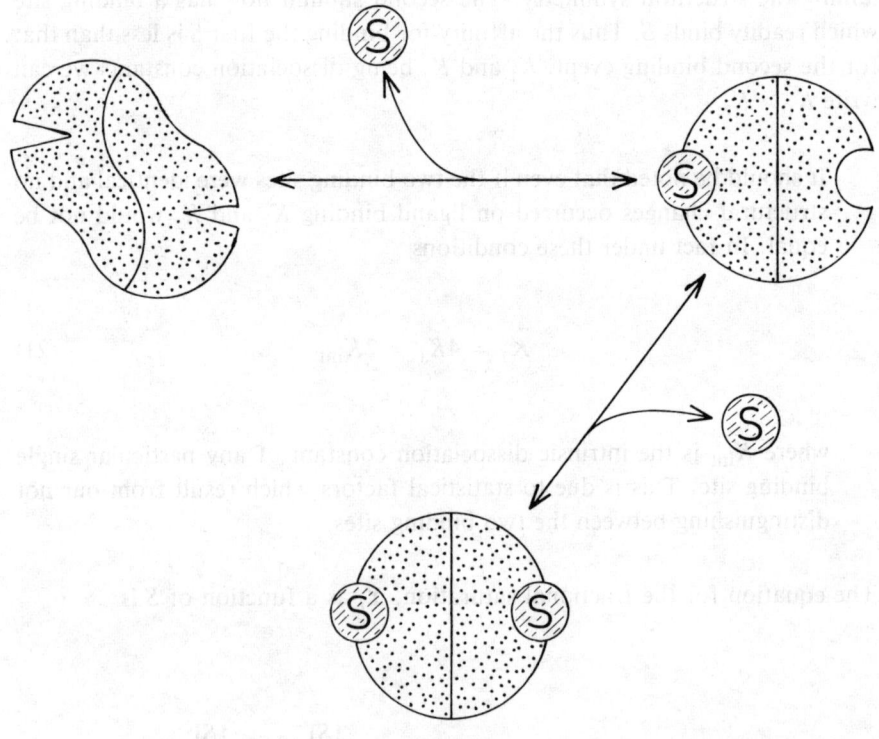

Fig. 4-9 Cooperative substrate binding to a dimeric protein.

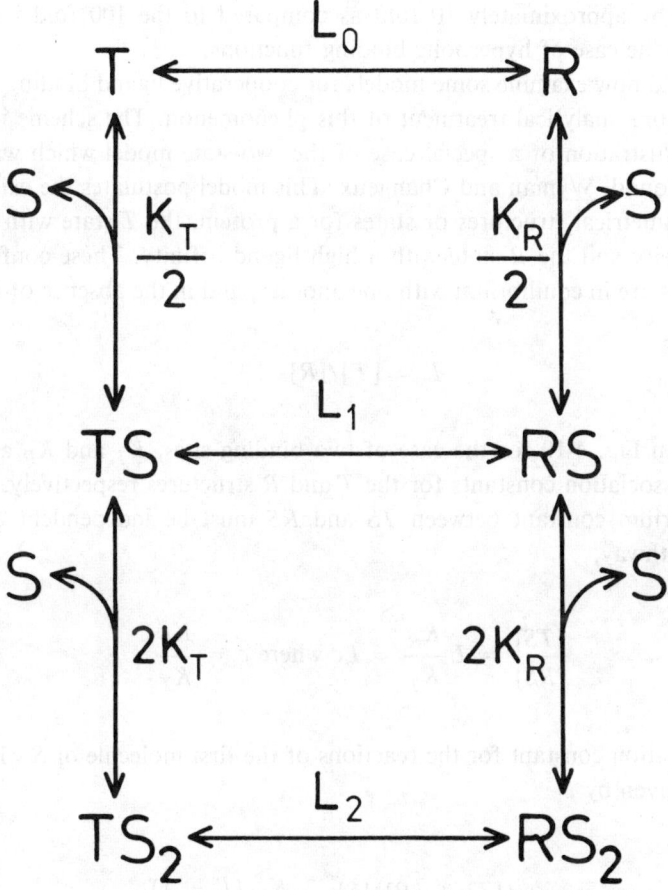

Fig. 4-10 Schematic representation of the two-state model for cooperative substrate binding.

which is a hyperbolic binding function. However, if $K_1 > K_2$, then in the limiting case $[PS] = 0$ and our function becomes

$$Y \approx \frac{[S]^2/K_1 K_2}{1 + [S]^2/K_1 K_2} \tag{24}$$

This is not a hyperbolic function. To increase Y from 0.1 to 0.9 one must increase S by approximately 10 fold as compared to the 100 fold increase required in the case of hyperbolic binding functions.

We should now examine some models for cooperative ligand binding which permit a more analytical treatment of this phenomenon. The scheme in Fig. 4-9 is an illustration of a special case of the two-state model which was put forth by Monod, Wyman and Changeux. This model postulates the existence of two symmetrical structures or states for a protein, the T state with a low ligand affinity and the R state with a high ligand affinity. These conformational states are in equilibrium with one another, and in the absence of ligand

$$L = [T]/[R].$$ (25)

As shown in Fig. 4-10 for the case of two binding sites, K_T and K_R are the intrinsic dissociation constants for the T and R structures respectively. Since the equilibrium constant between TS and RS must be independent of the reaction pathway,

$$\frac{[TS]}{[RS]} = L\frac{K_R}{K_T} = Lc \text{ where } c = \frac{K_R}{K_T}$$ (26)

The dissociation constant for the reactions of the first molecule of S with the protein is given by

$$K_1 = \frac{([T] + [R])[S]}{[TS] + [RS]} = \frac{K_R}{2}\frac{(L + 1)}{(Lc + 1)}$$ (27)

Likewise

$$\frac{[TS_2]}{[RS_2]} = Lc^2$$ (28)

and

$$K_2 = 2K_R\frac{(Lc + 1)}{(Lc^2 + 1)}$$ (29)

Note: For a protein of n subunits, the dissociation constants for the reaction of the ith ligand is given by

$$K_i = \frac{i}{n + 1 - i} K_R \frac{(Lc^{i-1} + 1)}{(Lc^i + 1)} \tag{30}$$

for $Lc^i \gg 1$

$$K_i = \frac{i}{n + 1 - i} K_T \tag{31}$$

for $Lc^{i-1} \ll 1$

$$K_i = \frac{i}{n + 1 - i} K_R \tag{32}$$

Let us consider a specific example. Suppose that $L = 10^2$, $C = 10^{-4}$ and $K_R - 10^{-3}$

then $\dfrac{[TS]}{[RS]} = 10^{-2}$ and $\dfrac{[TS_2]}{[RS_2]} = 10^{-6}$

This is then a situation similar to that pictured in Fig. 4-9. Solving for the dissociation constants we obtain $K_1 = 0.5 \times 10^{-1}$ and $K_2 = 2 \times 10^{-3} = 2K_R$. In the absence of ligand 99% of the protein molecules are in the T state. The preferential binding of ligand to the R state shifts this equilibrium so that 99% of the molecules with one ligand bound are in the R state, and the percentage approaches 100 when two ligands are bound.

It is quite clear that if we consider a protein with a greater number of binding sites, values of L and c can be chosen such that the protein remains in the T state when one ligand is bound but is in the R state for the final binding event. In such a case the affinity change induced by ligand binding will indeed be the change from K_T to K_R. This however is not possible when there are only two binding sites. With large numbers of subunits, the dependence of Y on $[S]$ can become progressively greater and can approach

$$Y = \frac{[S]^n}{K + [S]^n} \tag{33}$$

where n is less than or equal to the number of binding sites. If $n = 4$, then $[S]$ must change only 3 fold to go from $Y = 0.1$ to $Y = 0.9$.

Now let our subunits have, in addition to a substrate binding site, a site for the binding of an allosteric effector. Within the framework of the two-state model such an effector acts to alter the value of L. An allosteric inhibitor increases L by binding preferentially to the T state, while an allosteric activator binds preferentially to the R state decreasing L. The dependence of L on $[I]$, the inhibitor concentration, or on $[A]$, that of the activator, will of course be a complex one because of the cooperative nature of their binding to the protein. The ultimate effect of such cooperativity is to greatly increase the sensitivity of v to changes in both substrate and effector concentrations.

The two-state model is actually a very simple formulation, involving only three adjustable parameters, and its success in reproducing cooperative phenomena is impressive. However, its underlying assumptions, the requirement of absolute symmetry (or binding site identity) and the existence of only two conformational states, are extreme and unlikely to be generally true for all allosteric proteins exhibiting cooperative binding. There is an alternative model which postulates a sequential change in subunit conformation in response to ligand binding. Before examining this model it is worthwhile to explore another thermodynamic linkage which exists in systems exhibiting cooperative binding. This is the relationship between the free energy of subunit interaction and changes in ligand affinity. For this analysis we consider the reaction pathway in Fig. 4-11, again considering a dimeric protein with two binding sites for S. We are considering the reaction of the dimer with ligands and the dissociation of the dimer into monomeric units. Let K_m be the dissociation constant for the reaction of S with monomer. K_1 and K_2 are, as before, the dissociation constants for the ligand-dimer interaction. D_0, D_1 and D_2 are the dissociation constant for the dimer to monomer equilibrium when 0, 1 and 2 ligands are bound respectively. For convenience we will express K_1 and K_2 in terms of intrinsic dissociation constants to allow direct comparison with K_m

$$K_1 = \frac{1}{2}K_1' \qquad K_2 - 2K_2' \tag{34}$$

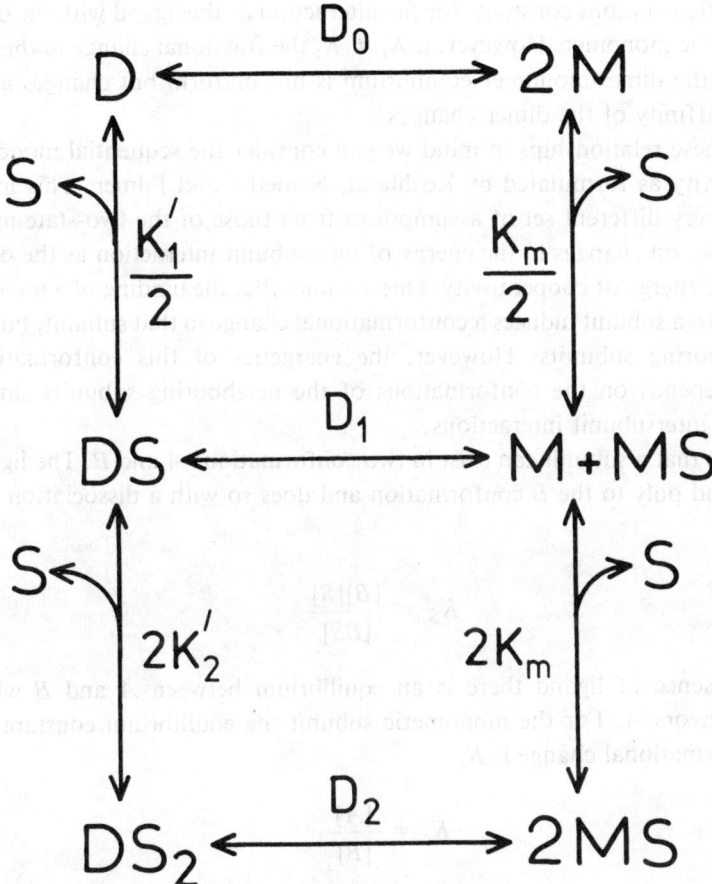

Fig. 4-11 Schematic illustration of the thermodynamic linkage between the dissociation of a dimeric protein to monomeric units and the binding of substrate molecules.

We can then write that

$$\frac{D_1}{D_0} = \frac{K_1'}{K_m} \quad \text{and} \quad \frac{D_2}{D_1} = \frac{K_2'}{K_m} \tag{35}$$

If $K_1' = K_2'$, then $D_1 = \sqrt{D_2 D_0}$ and the ratio D_1/D_0 equals the ratio of the intrinsic dissociations constants for the interaction of the ligand with the dimer and with the monomer. However, if $K_1' \neq K_2'$ the fractional change in the constant for the dimer-monomer equilibrium is not uniform but changes as the intrinsic affinity of the dimer changes.

With these relationships in mind we can consider the sequential model for cooperativity as formulated by Koshland, Nemethy and Filmer. This model makes a very different set of assumptions from those of the two-state model and focuses on changes in the energy of intersubunit interaction as the origin of the free energy of cooperativity. One assumes that the binding of a molecule of ligand to a subunit induces a conformational change in that subunit, but not in neighboring subunits. However, the energetics of this conformational change depends on the conformations of the neighboring subunits since it alters the intersubunit interactions.

Assume that a subunit can exist in two conformations A and B. The ligand, S, can bind only to the B conformation and does so with a dissociation constant K_S.

$$K_S = \frac{[B][S]}{[BS]} \tag{36}$$

In the absence of ligand there is an equilibrium between A and B which strongly favors A. For the monomeric subunit the equilibrium constant for this conformational change is K_c

$$K_c = \frac{[A]}{[B]} \tag{37}$$

The equilibrium constant for the dissociation of the dimer to monomeric subunits depends on the structural states of the subunit. We must therefore define these additional equilibrium constants

$$K_{AB} = \frac{[A][B]}{[AB]} \tag{38}$$

$$K_{AA} = \frac{[A]^2}{[A_2]} \tag{39}$$

$$K_{BB} = \frac{[B]^2}{[B_2]} \tag{40}$$

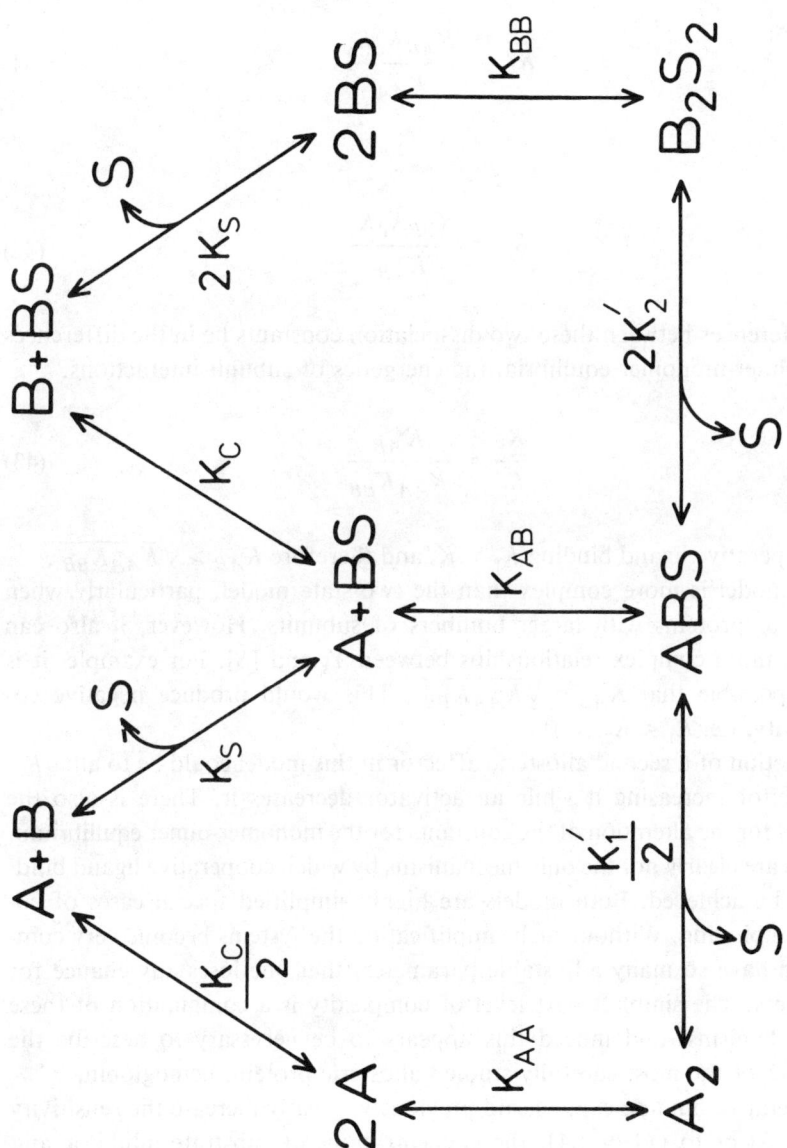

Fig. 4-12 Schematic illustration of the linkage relationships involved in the formulation of the sequential model for cooperative substrate binding.

One can now construct pathways for the binding of ligands to the dimer which are made up of only reactions with defined equilibrium constants, see Fig. 4-12. From this diagram we can see that

$$K_1' = \frac{K_{AB}K_cK_s}{K_{AA}} \tag{41}$$

and

$$K_2' = \frac{K_{BB}K_cK_s}{K_{AB}} \tag{42}$$

The differences between these two dissociation constants lie in the differences in the dimer-monomer equilibria, the energetics of subunit interactions.

$$\frac{K_1'}{K_2'} = \frac{K_{AB}^2}{K_{AA}K_{BB}} \tag{43}$$

For cooperative ligand binding $K_1' > K_2'$ and therefore $K_{AB} > \sqrt{K_{AA}K_{BB}}$.

This model is more complex than the two-state model, particularly when applied to proteins with larger numbers of subunits. However, it also can produce more complex relationships between Y_s and $[S]$. For example, it is clearly possible that $K_{AB} < \sqrt{K_{AA}K_{BB}}$. This would produce negative cooperativity, i.e. $K_1' < K_2'$.

The action of a second allosteric affector in this model could be to alter K_c, an inhibitor increasing it while an activator decreases it. There is also the potential for the alteration of the constants for the monomer-dimer equilibrium.

These are clearly not the only mechanisms by which cooperative ligand binding can be achieved. Both models are highly simplified special cases of cooperative binding. Without such simplification the systems become very complex and have so many adjustable parameters that one loses any chance for uniqueness. The simplest next level of complexity is a combination of these two mechanisms, and indeed this appears to be necessary to describe the properties of the most carefully studied allosteric protein, hemoglobin.

Cooperative binding can, as said previously, greatly increase the sensitivity of v to $[S]$ or to $[I]$ or $[A]$, the concentrations of substrate inhibitor and activator molecules. Combined with the complete flexibility that allosteric mechanisms allow in the choice of which substances will act as inhibitors or

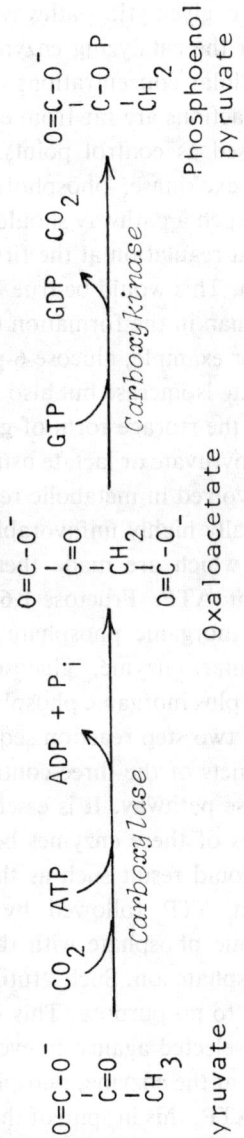

Fig. 4-13 The conversion of pyruvate to phosphoenol pyruvate by carboxylase and carboxykinase. This process requires the conversion of one ATP and one GTP to ADP and GDP respectively. The hydrolysis of GTP, guanosine triphosphate, is energetically equivalent to that of ATP.

activators, the potential for control of enzyme activity beyond that obtainable from classical inhibition is clear.

Let us now return to glycolysis, this time to examine the control of this metabolic pathway. Three steps in the glycolytic pathway are subject to control, i.e. variation in the activity of the catalyzing enzyme. They can be identified by examining the normal cellular concentrations of the glycolytic intermediates and determining which reactions are far from equilibrium since only such reactions can be usefully used as control points. The glycolytic enzymes which can be regulated are hexokinase, phosphofructokinase and pyruvate kinase. One can first ask why such a pathway should be regulated at more than one point. It would seem that regulation at the first step would be adequate and the most logical approach. This would be true if the intermediates in the pathway had no other role than in the formation of pyruvate and ATP. However, that is not the case. For example, glucose-6-phosphate is the substrate not merely for hexosephosphate isomerase but also for the enzymes that catalyse the synthesis of glycogen, the storage form of glucose. In addition organisms synthesize glucose from pyruvate or lactate using the glycolytic enzymes with the exception of those involved in metabolic regulation. Those three control steps, which are energetically highly unfavorable in reverse are bypassed by other reaction pathways which are made thermodynamically favorable by the loss or investment of ATP. Fructose-1,6-diphosphate is converted to fructose-6-phoshate plus inorganic phosphate by the enzyme fructose-1,6-diphosphatase and a similar enzyme, glucose-6-phosphatase converts glucose-6-phosphate to glucose plus inorganic phosphate. Pyruvate is converted to phosphoenolpyruvate by a two step reaction sequence as shown in Fig. 4-13. Therefore, all of the products of the three control enzymes are substrates for the enzymes of the reverse pathway. It is essential to the well being of the organism that the activities of these enzymes be carefully controlled. Without such control, cycles would result such as the formation of glucose-6-phosphate from glucose and ATP followed by its immediate hydrolysis to form glucose and inorganic phosphate with the net effect of hydrolyzing ATP to form ADP and phosphate ion. Such "futile cycles" could dissipate the organism's supply of ATP to no purpose. This would surely be life threatening, and would be strongly selected against in evolution.

The major control of glycolysis occurs at the enzyme, phosphofructokinase. This is an enzyme which is inhibited by ATP, this in spite of the fact that ATP is a substrate for the enzyme. The inhibitory effect of ATP results from binding not at the catalytic sites but at specific inhibitor binding sites on this allosteric enzyme. Citrate, another ultimate product of glycolysis also acts as an allosteric inhibitor of this enzyme. At the same time the enzyme is activated by AMP. This complex pattern of control serves to make the enzyme sensitive to two aspects of the state of the cell upon which this pathway has an effect. Obviously one is the concentration of ATP. The second, which is perhaps not obvious from what we have discussed thus far, is the overall concentration of the intermediates of the Krebs, citric acid cycle. The intermediates are important, not merely in promoting the oxidation of acetate with the ultimate formation of ATP, but also as precursors for a number of important biosynthetic reactions. Given that the enzyme is inhibited by ATP, one might ask what is accomplished by its being activated by AMP. The answer is a greatly increased sensitivity to the ATP state of the cell. In the cell the total concentration of phosphorylated adenosine is very nearly constant and equal to the sum of ATP, ADP and AMP

$$[AXP] = [ATP] + [ADP] + [AMP]$$

An enzyme catalyzed equilibrium exists among these three compounds according to the reaction

$$2ADP \rightleftharpoons ATP + AMP$$

such that

$$K = \frac{[ATP][AMP]}{[ADP]^2} \approx 1$$

Under normal circumstances most of AXP is ATP; the normal range is 80–95%. Under these circumstances the absolute concentration of ATP varies by only fractional amounts. The control of ATP concentration within such narrow limits is difficult to achieve if one relies only on the differential binding of ATP to allosteric regulating sites on an enzyme. On the other hand, ADP and AMP concentrations vary by much greater factors. When ATP remains fairly high, AMP is smaller than ADP since

$$\frac{[AMP]}{[ADP]} = \frac{[ADP]}{[ATP]}$$

However, it varies by a greater factor since it depends on the square of [ADP]

$$[AMP] = \frac{[ADP]^2}{[ATP]}$$

Therefore, designing the enzyme to be activated by AMP, or rather to have its inhibition by ATP reversed by AMP, is an ideal way to increase its sensitivity to small changes in ATP concentration.

In higher organisms this enzyme is also activated by another compound, fructose-2, 6-biphosphate, which is formed by another enzyme that can be controlled by factors outside the cell through a process which results in covalent modification of this enzyme. This is an example of hormonal control of enzyme activity which we will discuss later in the context of glycogen biosynthesis and utilization.

The control of the enzyme, hexokinase, appears to be much less elaborate than that of phosphofructokinase. This enzyme is strongly inhibited by its product, glucose-6-phosphate, again by an allosteric mechanism. Therefore, significant accumulation of glucose-6-phosphate, either because of inhibition of phosphofructokinase, the activity of fructose-1,6-diphosphatase, or the breakdown of glycogen blocks further formation from glucose. Pyruvate kinase is allosterically inhibited by ATP.

Control of Enzymatic Activity by Covalent Modification

There is one more mechanism for controlling enzymatic activity to which we have alluded but which we should now consider in more detail. This is control by covalent modification. In this regard we will consider the control of the synthesis and breakdown of glycogen.

Glycogen is the storage form of glucose. It is composed of glucose residues bound together primarily by α (1-4) glycosidic linkages, as shown in Fig. 4-14. These linkages have a large negative free energy of hydrolysis, and thus the formation of glycogen even from glucose-6-phosphate requires the investment of an ATP in order to be energetically favorable. This synthesis is achieved by the pathway shown in Fig. 4-15. The reversal of this biosynthetic pathway to form glucose-6-phosphate from glycogen involves phosphorolysis of this glycosidic linkage by the enzyme glycogen phosphorylase. Both this enzyme and the enzyme, glycogen synthase, are subject to control by phosphorylation. If both of these enzymes are active there will be a continuous formation and phosphorolysis of glycogen with a futile expenditure of ATP. However, these enzymes can exist in an active and in an inactive form. Glycogen synthase can

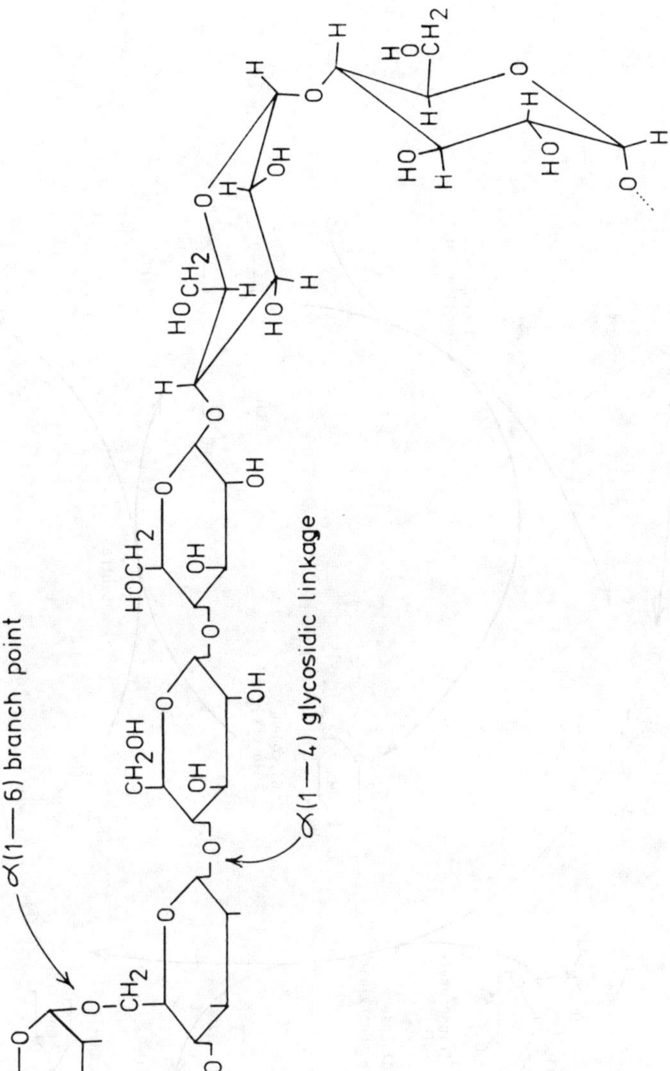

Fig. 4-14 Glycogen

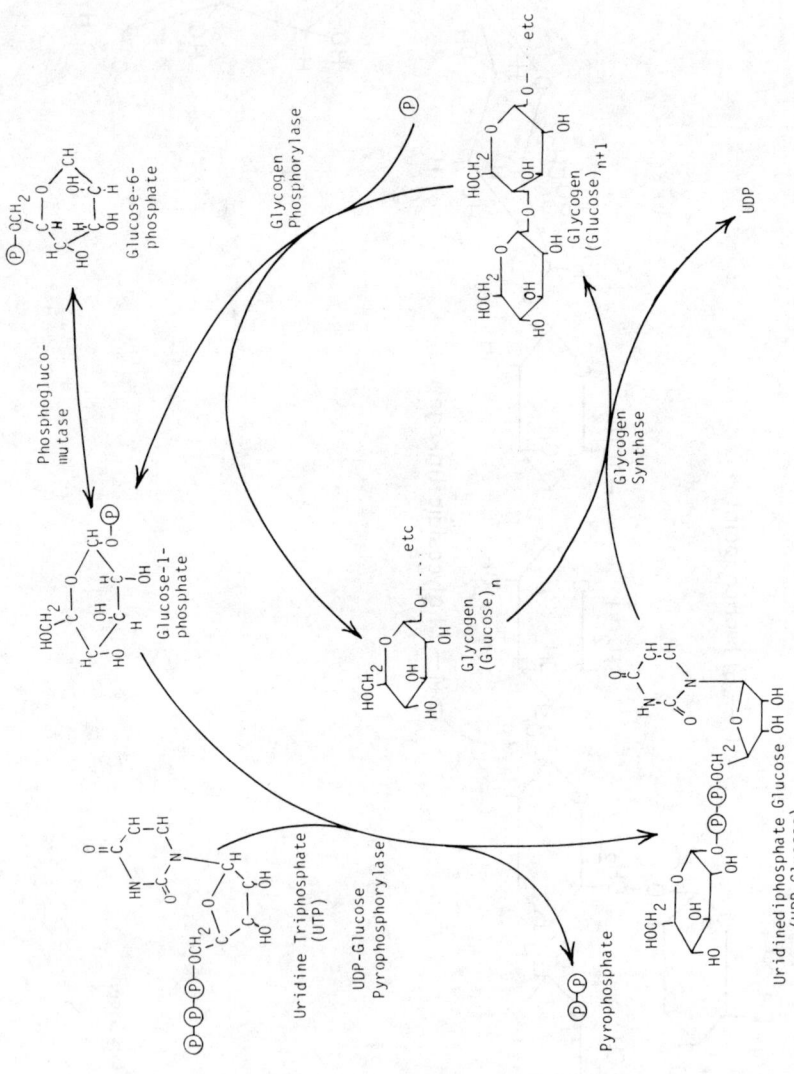

Fig. 4-15 Pathways for the synthesis and breakdown of glycogen.

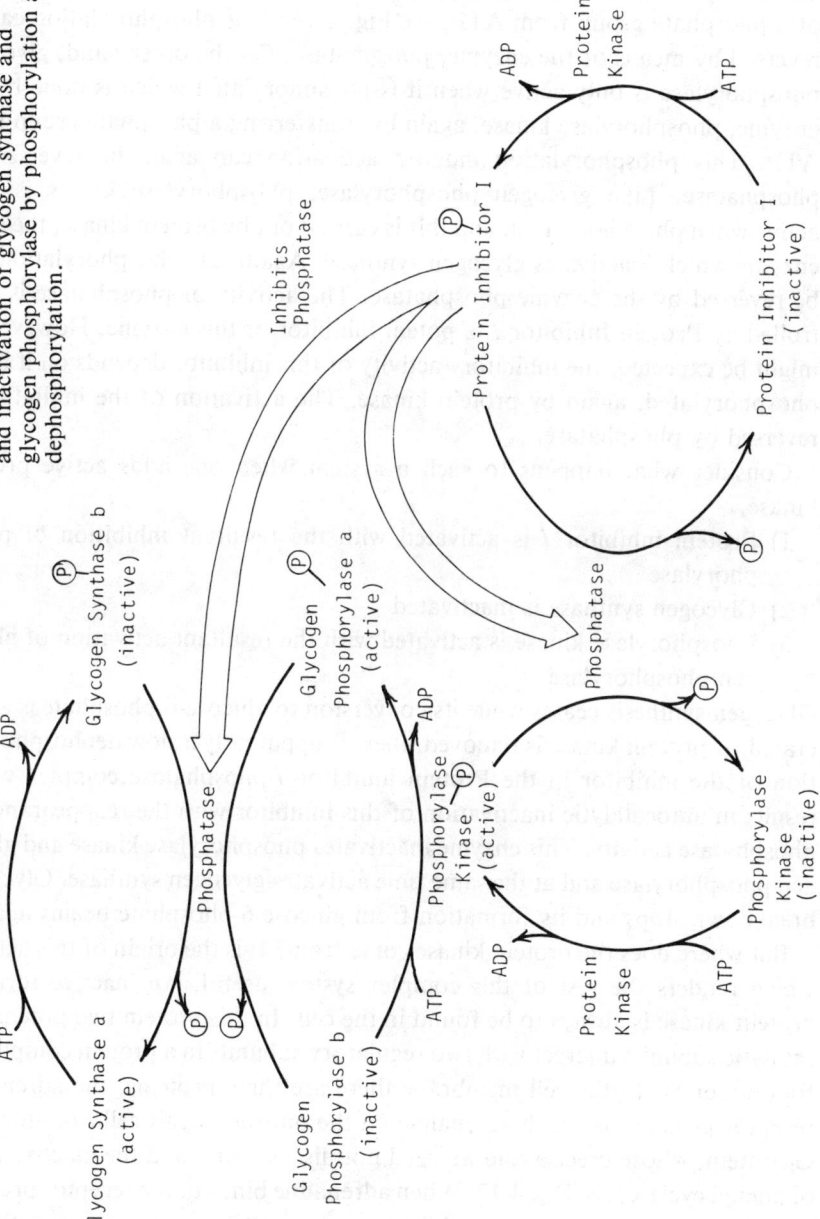

Fig. 4-16 Diagram of the enzymatic activation and inactivation of glycogen synthase and glycogen phosphorylase by phosphorylation and dephosphorylation.

be inactivated by the phosphorylation of a specific amino acid residue. This phosphorylation is carried out by the enzyme, protein kinase, by the transfer of a phosphate group from ATP, see Fig. 4-16. This phosphorylation can be reversed by means of the enzyme, phosphatase. On the other hand, glycogen phosphorylase is only active when it is phosphorylated which is done by the enzyme, phosphorylase kinase, again by transferring a phosphate group from ATP. This phosphorylation induced activation can again be reversed by phosphatase. Like glycogen phosphorylase, phosphorylase kinase is only active when phosphorylated, and this is carried out by protein kinase, the same enzyme which inactivates glycogen synthase. Again this phosphorylation can be reversed by the enzyme phosphatase. The activity of phosphatase is controlled by Protein Inhibitor *I*, a potent inhibitor of this enzyme. However, as might be expected, the inhibitory activity of this inhibitor depends on it being phosphorylated, again by protein kinase. The activation of the inhibition is reversed by phosphatase.

Consider what happens to such a system when one adds active protein kinase.

1) Protein inhibitor *I* is activated with the resultant inhibition of phosphorylase
2) Glycogen synthase is inactivated
3) Phosphorylase kinase is activated with the resultant activation of glycogen phosphorylase

Glycogen synthesis ceases while its conversion to glucose-6-phosphate is accelerated. If protein kinase is removed, there is apparently a slow dephosphorylation of the inhibitor in the Protein inhibitor *I*-phosphatase complex which results in autocatalytic inactivation of this inhibitor with the reappearance of phosphatase activity. This enzyme inactivates phosphorylase kinase and glycogen phosphorylase and at the same time activates glycogen synthase. Glycogen breakdown stops and its formation from glucose-6-phosphate begins again.

But where does the protein kinase come from? It is the origin of this activity which renders the rest of this complex system useful. An inactive form of protein kinase is always to be found in the cell. In this protein two potentially catalytic subunits interact with two regulatory subunits in a protein complex of four subunits. In the cell membrane there are three proteins, the adrenaline receptor protein, to which adrenaline on the outside of this cell can bind; the G-protein, whose precise role we need not discuss here; and the inactive form of adenyl cyclase, see Fig. 4-17. When adrenaline binds to the receptor protein, it alters the conformation of this protein so that it, in concert with the G-protein, can bind to and activate adenyl cyclase. This is a straight forward

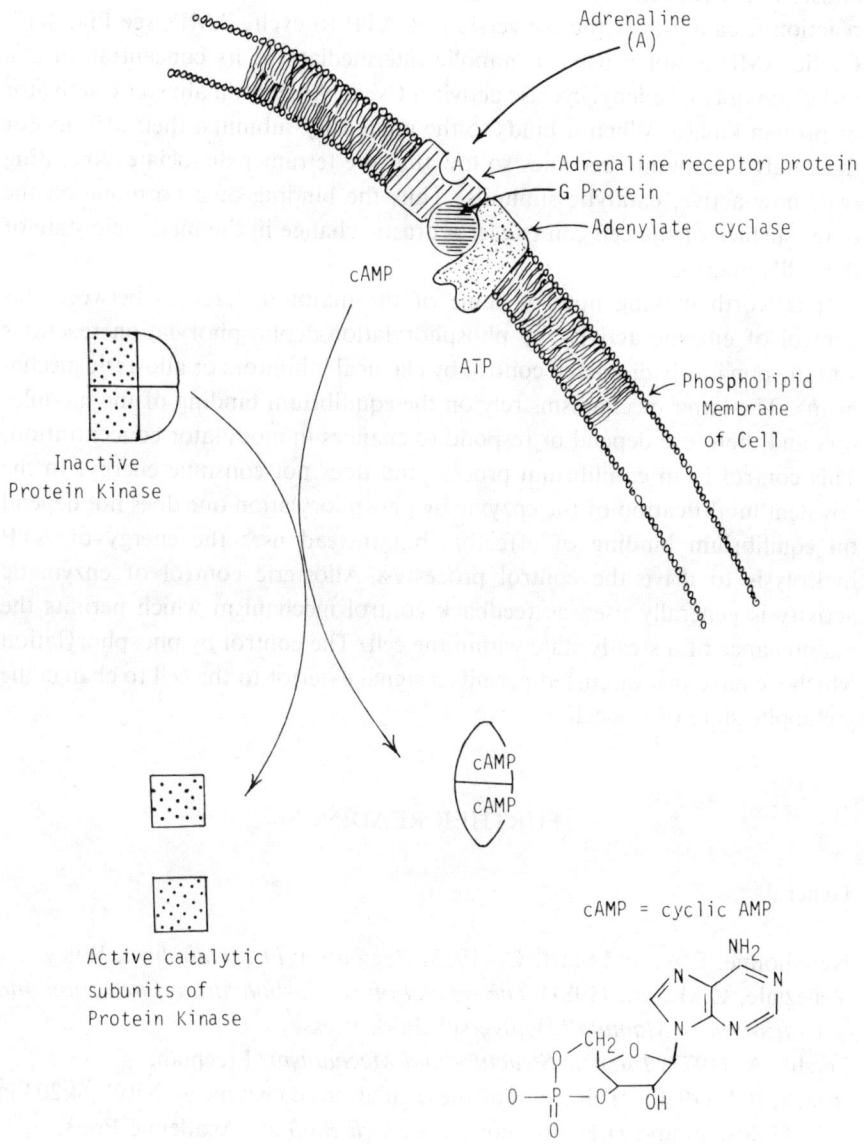

Adrenaline
(A)

Adrenaline receptor protein

G Protein

Adenylate cyclase

cAMP

ATP

Phospholipid
Membrane
of Cell

Inactive
Protein Kinase

cAMP
cAMP

cAMP = cyclic AMP

Active catalytic
subunits of
Protein Kinase

Fig. 4-17 Diagram of the activation of protein kinase. This enzyme is activated by cyclic AMP which is produced by adenyl cyclase, and enzyme whose activity is modulated by adrenaline binding to a receptor protein on the exterior at the cell membrane.

allosteric activation. The active site of adenyl cyclase is inside the cell. The reaction it catalyses is the conversion of ATP to cyclic AMP, see Fig. 4-17. Cyclic AMP is not a usual metabolic intermediate so its concentration is a reflection only of adenyl cyclase activity. Cyclic AMP is an allosteric activator of protein kinase. When it binds to the regulatory subunits, their affinity for the catalytic subunits becomes so low that the tetramer dissociates liberating two, now active, catalytic subunits. Thus the binding of a hormone on the outer surface of the cell can trigger a drastic change in the metabolic state of the cell's interior.

It is worth making note of some of the major differences between this control of enzyme activity by phosphorylation-dephosphorylation reactions and the previously discussed control by classical inhibitors or allosteric mechanisms. The latter mechanisms rely on the equilibrium binding of the modulators and therefore depend or respond to changes in modulator concentration. This control is an equilibrium process and does not consume energy. In the covalent modification of the enzyme by phosphorylation one does not depend on equilibrium binding of effectors but instead uses the energy of ATP hydrolysis to drive the control processes. Allosteric control of enzymatic activity is generally used as feedback control mechanism which permits the maintenance of a steady state within the cell. The control by phosphorylation which we have just discussed permits a signal exterior to the cell to change the metabolic state of the cell.

FURTHER READINGS

General

Newsholme, E.A. and Start, C. (1973) *Regulation in Metabolism*, Wiley.

Veneziale, C.M., ed. (1981) *The regulation of Carbohydrate Formation and Utilization in Mammals*, University Park Press.

Fersht, A. (1977) *Enzyme Structure and Mechanism*, Freeman.

Roach, P.J. (1980) "Principles of the regulation of enzyme activity", 4:203 in L. Goldstein and D.H. Prescott, eds., *Cell Biology*, Academic Press.

Specific

Monod, J., Wyman, J. and Changeux, J.P. "On the nature of allosteric transitions", (1965) *J. Mol. Biol.* **12**:88–118.

Koshland, D.E. Jr., Nemethy, G. and Filmer, D. "Comparison of experimental binding data and theoretical models in proteins containing subunits", (1966) *Biochemistry* **5**:365–385.

Ross, E.M. and Gilman, A.C. "Biochemical properties of hormone sensitive adenylate cyclase", (1980) *Annu. Rev. Biochem.* **49**:533.

CHAPTER 5

MEMBRANES, THE BOUNDARIES OF LIVING SYSTEMS

He will not go behind his father's saying,
And he likes having thought of it so well
He says again, "Good fences make good neighbors."

Robert Frost
"Mending Wall"

One of the characteristic features of living systems is that they have distinct boundaries which separate them from their environment. These boundaries are effective diffusion barriers permitting the establishment of differences in solute concentration between the interior of the organism and the outside world.

Biological membranes are composed predominantly of lipids and proteins. The lipids form a bilayer structure into which the proteins are inserted or dissolved. The stability of the bilayer results from the amphipathic nature of the lipids of which it is composed. Although there are many different lipid molecules, almost all of them have in common a charged or polar moiety from which two long hydrocarbon chains extend. Typical of such molecules are Phosphatidyl serine and Phosphatidyl ethanolamine whose structures are

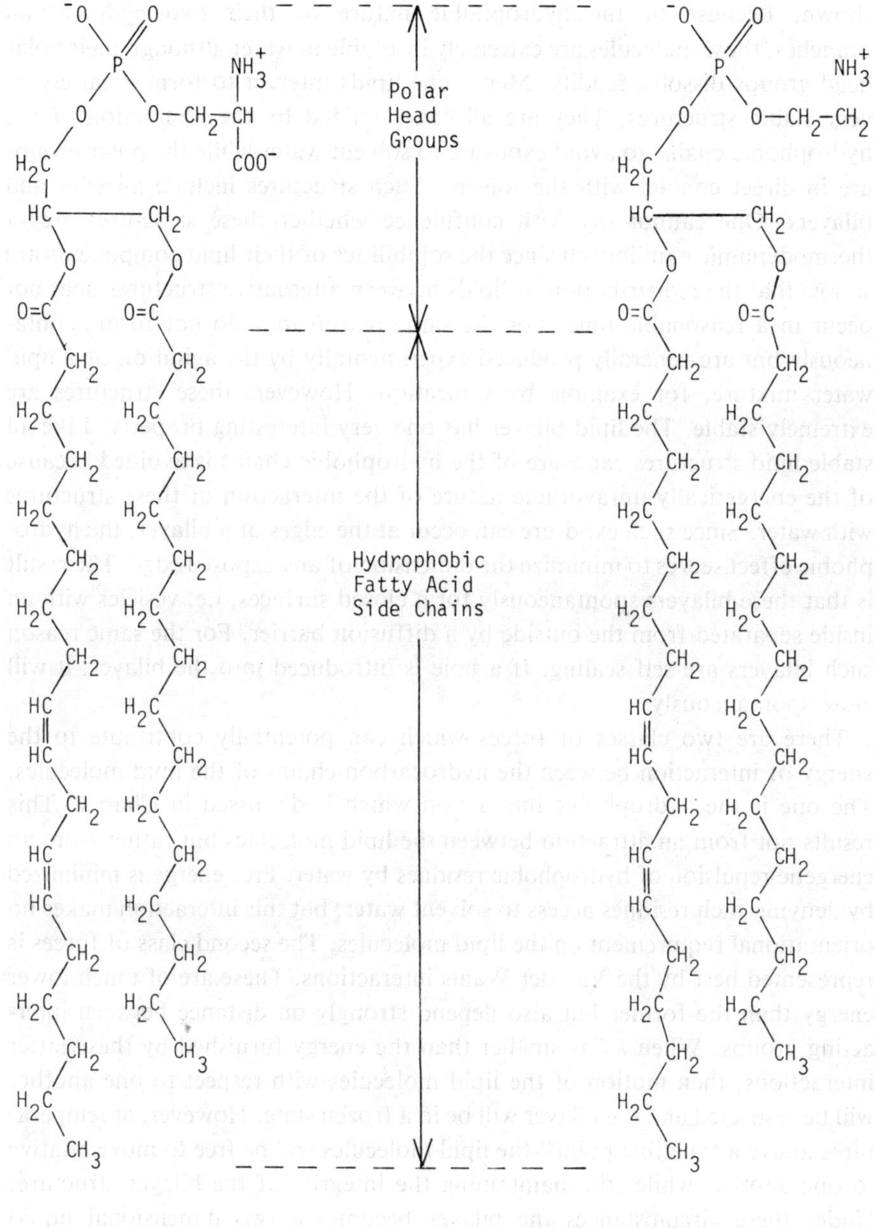

Phosphatidyl Serine

Phosphatidyl Ethanolamine

Fig. 5-1 Two examples of phospholipids.

shown. Because of the hydrophobic nature of their two hydrocarbon branches, these molecules are extremely insoluble in water although their polar head groups dissolve readily. Membrane lipids interact to form a variety of very stable structures. They are all characterized by the interaction of the hydrophobic chains to avoid exposure to solvent water while the polar groups are in direct contact with the solvent. Such structures include micelles and bilayers. One cannot say with confidence whether these structures are at thermodynamic equilibrium since the solubilities of their lipid components are so low that the redistribution of lipids between alternative structures does not occur in a reasonable time. For the same reason they do not form spontaneously but are generally produced experimentally by the agitation of a lipid water mixture, for example by sonication. However, these structures are extremely stable. The lipid bilayer has one very interesting property. Like all stable lipid structures exposure of the hydrophobic chains is avoided because of the energetically unfavorable nature of the interaction of these structures with water. Since such exposure can occur at the edges of a bilayer, the hydrophobic effect serves to minimize the dimension of any exposed edge. The result is that these bilayers spontaneously form closed surfaces, i.e. vesicles with an inside separated from the outside by a diffusion barrier. For the same reason such bilayers are self sealing. If a hole is introduced into the bilayer, it will close spontaneously.

There are two classes of forces which can potentially contribute to the energy of interaction between the hydrocarbon chains of the lipid molecules. The one is the hydrophobic interaction which is discussed in Chap 7. This results not from an attraction between the lipid molecules but rather from an energetic repulsion of hydrophobic residues by water. Free energy is minimized by denying such residues access to solvent water, but this interaction makes no orientational requirement on the lipid molecules. The second class of forces is represented best by the Van der Waals interactions. These are of much lower energy than the former but also depend strongly on distance between interacting groups. When kT is smaller than the energy furnished by these latter interactions, then motion of the lipid molecules with respect to one another will be restricted and the bilayer will be in a frozen state. However, at temperatures above a "melting point" the lipid molecules will be free to move relative to one another while still maintaining the integrity of the bilayer structure. Under these circumstances the bilayer becomes a two dimensional liquid phase. The melting properties of the membrane mimic those of three dimensional phases. A bilayer composed of a single type of lipid molecule will exhibit a sharp melting point. Introduction of a second lipid solute with lower melting

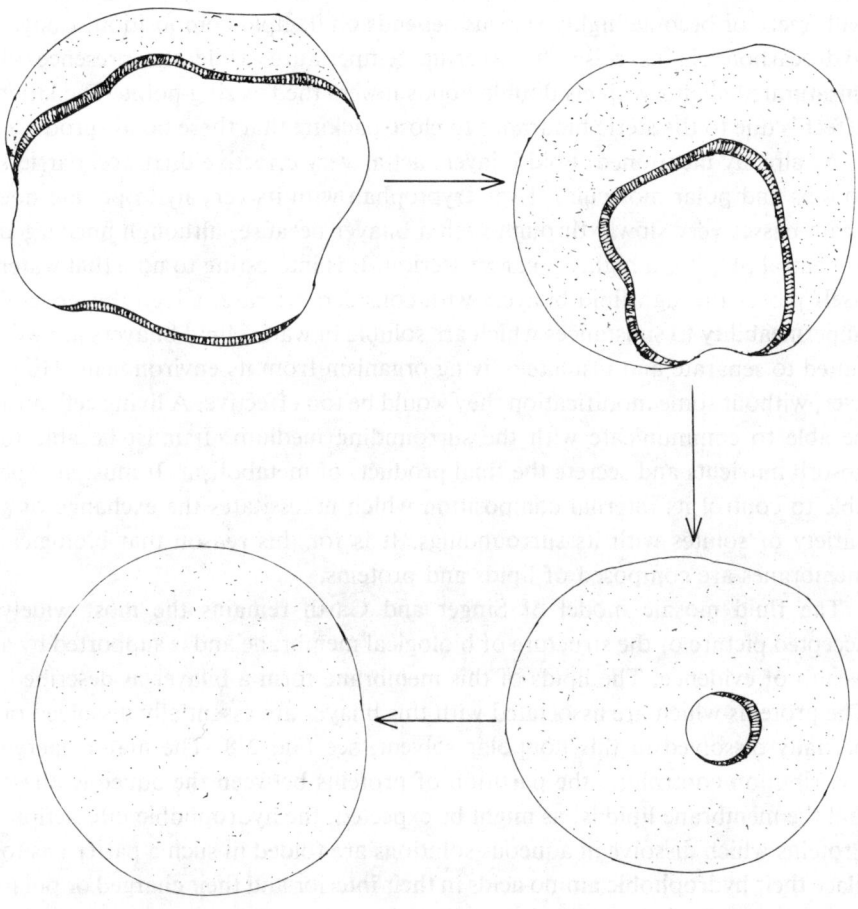

Fig. 5-2 The spontaneous closing of a sheet lipid bilayer to form a closed surface or vesicle.

point will depress the melting point of the solvent lipid. Furthermore, the freezing of the solvent lipid will effect a physical separation between solvent and solute. A bilayer composed of a complex mixture of lipids will not exhibit a sharp melting point. Instead there will be a range of temperature over which the viscosity of the bilayer increases rapidly with decreasing temperature until the membrane becomes essentially solid. The temperature at which a bilayer will freeze or become highly viscous depends on its lipid composition. Longer hydrocarbon chains raise the freezing temperature while the presence of unsaturated chains with cis double bonds lowers the freezing point. The latter effect is due to the steric hindrance to close packing that these bonds produce.

As already mentioned, lipid bilayers act as very effective diffusion barriers to ions and polar molecules. Even tryptophan with its very hydrophobic side chain passes very slowly through such a bilayer because, although uncharged at neutral pH, it is a highly polar zwitterion. It is interesting to note that water itself passes through lipid bilayers with considerable ease. Given this general impermeability to substances which are soluble in water, lipid bilayers are well suited to separate and insulate a living organism from its environment. However, without some modification they would be too effective. A living cell must be able to communicate with the surrounding medium. It must be able to absorb nutrients and secrete the final products of metabolism. It must also be able to control its internal composition which necessitates the exchange of a variety of solutes with its surroundings. It is for this reason that biological membranes are composed of lipids and proteins.

The fluid mosaic model of Singer and Garth remains the most widely accepted picture of the structure of biological membrane and is supported by a wealth of evidence. The lipids of this membrane form a bilayer as described. The proteins which are associated with this bilayer are essentially dissolved or partially dissolved in this nonpolar solvent, see Fig. 2-8. The major energy contribution controlling the partition of proteins between the aqueous phase and the membrane lipid is, as might be expected, the hydrophobic interaction. Proteins which dissolve in aqueous solutions are folded in such a pattern as to place their hydrophobic amino acids in their interior and their charged or polar amino acid side chains on their surface, where they are in contact with water. A protein which has all or part of its surface occupied by hydrophobic residues will form a stable association with a lipid bilayer. Its orientation with respect to the bilayer will depend on the pattern of hydrophobic and hydrophilic regions on its surface, since it will tend to be positioned so as to minimize contact between hydrophobic residues and water at the same time minimizing the insertion of polar or charged residues into the nonpolar interior of the membrane. Again, since the hydrophobic interaction makes no specifications as to the distance between or precise orientations of the nonpolar residues involved, the protein is free to move about in the two dimensional solution of the membrane. However, since the bilayer is an effective diffusion barrier to charged or polar molecules, only lateral motion is permitted. The protein cannot rotate through the membrane since its hydrophilic surface cannot enter the interior of the membrane. For the same reason lipid molecules cannot easily rotate from one side of the membrane to another. For this reason such membranes can be, and in fact generally are, asymmetric.

This potential for asymmetry emphasizes the stability of such systems even when they are clearly not at equilibrium. They represent deep local energy minima, but uncatalyzed redistribution is extremely slow. Membrane proteins, like lipids, are in general extremely insoluble in aqueous solution and to avoid this necessity they are inserted into the membrane system of the cell during synthesis.

The impermeability of biological membranes creates a new set of processes by which living systems can approach equilibrium. Transport processes, although not chemical reactions, are associated with defined free energy changes and can be catalyzed in the same sense as a chemical modification. Consider two compartments, 1 and 2, containing n_1 and n_2 moles of solute at concentrations c_1 and c_2.

If we transfer Δn moles of component i from compartment 1 to compartment 2, where $\Delta n \ll n_1, n_2$, then the free energy change in compartment 1 is

$$\Delta G_1 = -\Delta n (\mu_i)_1 \tag{1}$$

where $(\mu_i)_1$ is the chemical potential of component i in this compartment. At the same time the free energy change in compartment 2 is given by

$$\Delta G_2 = \Delta n (\mu_i)_2 \tag{2}$$

The free energy change in the entire system is given by

$$\Delta G = \Delta G_1 + \Delta G_2 = \Delta n \{ (\mu_i)_2 - (\mu_i)_1 \} \tag{3}$$

For dilute solutions of uncharged solutes

$$\mu_i = RT \ln c_i + \mu_i^0 \tag{4}$$

Therefore

$$\frac{\Delta G}{\Delta n} = RT \ln \frac{c_1}{c_2} \tag{5}$$

Let us assume that these two compartments are separated by a membrane, the lipid bilayer portion of which has zero permeability to our solute. We can imagine two ways in which proteins could catalyze the flow of solute from the compartment with the higher concentration to that with the lower. One would be the existence of channels in the protein through which the solute can diffuse. In such passive diffusion the flux is given by

$$\frac{dn}{dt} = \phi = -DA\frac{dc}{dx} \tag{6}$$

where D is the diffusion coefficient of the solute, A the area through which it is diffusing and dc/dx is the concentration gradient. For the system with which we are dealing dc/dx at steady state can be approximated by

$$\frac{dc}{dx} = \frac{c_1 - c_2}{r} = \frac{\Delta c}{r} \tag{7}$$

where r is the thickness of the membrane. That is, the flux of solute across the membrane will be proportional to Δc.

$$\frac{dn}{dt} = -DA\frac{\Delta c}{r} = -K\frac{\Delta c}{r} \tag{8}$$

This is passive diffusion.

The second mechanism is transport via a carrier protein. In this mechanism a solute molecule binds to a carrier protein which moves to the opposite side of the membrane where it releases the solute

$$S^{(1)} + C^{(1)} \underset{k_{-1}}{\overset{k_1}{\rightleftharpoons}} SC^{(1)} \underset{k_{-2}}{\overset{k_2}{\rightleftharpoons}} SC^{(2)} \underset{k_{-3}}{\overset{k_3}{\rightleftharpoons}} S^{(2)} + C^{(2)} \tag{9}$$

The general solution to the time course of three sequential, reversible reactions is complex, but for our purpose this system can be greatly simplified. Assume that the rate limiting step in this system is the movement of the carrier from one side of the membrane to another. Then we can reasonably assume, to a first approximation, that reactions 1 and 3 are always essentially at equilibrium and are identical in the sense that

$$k_1 = k_{-3} \text{ and } k_{-1} = k_3 \tag{10}$$

if this is so then $k_2 = k_{-2}$ since at equilibrium

$$[S^{(1)}]_{eq} = [S^{(2)}]_{eq} \tag{11}$$

Initially let us assume that $S^{(2)} = 0$. Then the flux of solute from compartment 1 to compartment will be proportioned to the number of carrier proteins in the membrane, n_{cp}, the fractional saturation of these carriers with S in compartment 1, Y_1, and the first order transfer constant, k_2

$$\frac{dn}{dt} = -Y_1 n_{cp} k_2 = \frac{[S^{(1)}]}{\dfrac{k_{-1}}{k_1} + [S^{(1)}]} n_{cp} k_2 \tag{12}$$

Obviously this is a saturable function of the same form obtained in the Michaelis-Menton formulation of enzyme catalysis of chemical reactions with $n_{cp} k_2$ being the equivalent of V_{max}.

   Note: The situation is in fact somewhat more complex than described. At most only 1/2 of the carrier proteins will be available for interaction at any one time in one side of the membrane. However, this factor is included in the constant k_2.

But what if $S^{(2)} > 0$? There are two ways to envision this. Perhaps the easiest is to realize that there will now be a flux of S from compartment 2 to 1 which will be equal to

$$\left(\frac{dn}{dt}\right)_2 = Y_2 n_{cp} k_{-2} = Y_2 n_{cp} k_2$$

$$= \frac{[S^{(2)}]}{\dfrac{k_3}{k_{-3}} + [S^{(2)}]} n_{cp} k_2 = \frac{[S^{(2)}]}{\dfrac{k_{-1}}{k_1} + [S^{(2)}]} n_{cp} k_2 \tag{13}$$

Since this is in the opposite direction to the net flux we are considering it will reduce it and we obtain

$$\frac{dn}{dt} = n_{cp}k_2 \left\{ \frac{[S^{(1)}]}{K_S + [S^{(1)}]} - \frac{[S^{(2)}]}{K_S + [S^{(2)}]} \right\} \tag{14}$$

$$= (\frac{dn}{dt})_{max} (Y_1 - Y_2)$$

With such a carrier there are two situations in which dn/dt goes to zero. One is obviously when $[S^{(1)}] = [S^{(2)}]$ and the system is at equilibrium. The other is when $Y_1 = Y_2 = 1$, that is when the carrier is saturated on both sides of the membrane even if $[S^{(1)}] \neq [S^{(2)}]$.

One can legitimately inquire about the mechanisms by which such transport proteins function. The existence of channels is now well documented. Multi-subunit complexes in which the subunits surround a central channel have been observed. Some of these oligomeric structures appear to be able to change their conformation in response to allosteric effectors in such a way as to open or close the channel depending on the signal received.

The mechanism by which carriers function is still a matter of speculation. Because of the saturable nature of this transport mechanism, a channel through a transmembrane protein is not an adequate explanation. Instead one requires a binding site. The question is how we can design such a site so that a molecule, when bound to it, can have access to either side to the membrane. This could be achieved by a carrier molecule that can rotate through the membrane (see Fig. 5-3), but as already discussed such motion by membrane proteins appears to be improbable. Another possible mechanism is an internal binding site in a protein which can exist in two conformational states, one in which the binding site has access to the interior of the cell and the other in which it has access to the exterior. Such a carrier protein might well consist of more than one subunit, as suggested in Fig. 5-4. The similarity to allosteric enzymes is obvious. This model has a conceptual advantage over the former in that it can be modified rather trivially in order to account for energy dependent transport, or membrane pumps, which we shall discuss shortly.

In the simple formulation of carrier transport which we have considered, the ligand affinity of the carrier remained identical on the two sides of the membrane. However, when we invoke a conformational change in the carrier protein as part of the mechanism, this simplifying assumption is clearly not necessarily true. If the affinity of the carriers does change, then the thermodynamic constraints on the system require other asymmetries. This can be seen by considering our system at equilibrium to be a closed cycle of reactions,

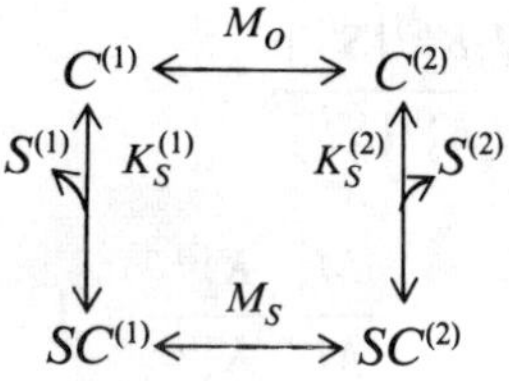

$$
\begin{array}{ccc}
C^{(1)} & \xleftrightarrow{\;M_O\;} & C^{(2)} \\[2pt]
S^{(1)}\Big\updownarrow\ K_S^{(1)} & & K_S^{(2)}\ \Big\updownarrow S^{(2)} \\[2pt]
SC^{(1)} & \xleftrightarrow{\;M_S\;} & SC^{(2)}
\end{array}
$$

where M_O, M_S, $K_S^{(1)}$ and $K_S^{(2)}$ are equilibrium constants and at equilibrium are given by

$$
M_O = \frac{[C^{(1)}]}{[C^{(2)}]}, \qquad M_S = \frac{[SC^{(1)}]}{[SC^{(2)}]}
$$

$$
K_S^{(1)} = \frac{[S^{(1)}][C^{(1)}]}{[SC^{(1)}]}, \qquad K_S^{(2)} = \frac{[S^{(2)}][C^{(2)}]}{[SC^{(2)}]}.
$$
(15)

The first law of thermodynamics requires that

$$
K_S^{(2)} M_O = K_S^{(1)} M_S
$$
(16)

or

$$
\frac{K_S^{(1)}}{K_S^{(2)}} = \frac{M_O}{M_S}
$$
(17)

Any asymmetry in the ligand affinity of the carrier requires a ligand dependence in the equilibrium between its conformational states. The importance of this can be seen if we consider the flux from side 1 to side 2,

$$
\phi_{1-2} = k_2 [SC^{(1)}]
$$
(18)

and that from side 2 to side 1,

$$
\phi_{2-1} = k_{-2} [SC^{(2)}]
$$
(19)

At equilibrium these must be equal or their ratio equal to unity.

$$\frac{(\phi_{1-2})_{eq}}{(\phi_{2-1})_{eq}} = 1 = \frac{k_2}{k_{-2}} \frac{[SC^{(1)}]}{[SC^{(2)}]} = \frac{k_2[S^{(1)}][C^{(1)}]K_S^{(2)}}{k_{-2}[S^{(2)}][C^{(2)}]K_S^{(1)}}$$

(20)

$$= \frac{M_o K_S^{(2)}[S^{(1)}]}{M_s K_S^{(1)}[S^{(2)}]}$$

Since
$$\frac{M_o}{M_1} \frac{K_S^{(2)}}{K_S^{(1)}} = 1 ,$$
(21)

at equilibrium $[S^{(1)}] = [S^{(2)}]$, as must be so. Such considerations may seem trivial, but these systems can easily become sufficiently complex that seemingly reasonable mechanisms for membrane phenomena are often found to violate these requirements.

Now that we have considered the process of transporting a single solute across a membrane, it must be understood that such a process can be coupled to other processes or chemical reactions by specific catalysis in a manner similar to that we considered when we discussed the enzymatic coupling of chemical reactions. Such coupling by linking an energetically favorable process to one that is unfavorable can permit us to use a concentration gradient to supply energy to another process or to use a chemical reaction as a source of energy in order to pump solute molecules across a membrane against a concentration gradient.

A trivial example of such coupling is the chemical modification of the solute on one side of the membrane, thereby continually reducing its concentration so that passive diffusion from the other side of the membrane is always energetically favorable. The conversion of glucose, G, to glucose-6-phosphate, G-6-P, within a cell is a perfect example of this which was discussed in Chapter 4. Glucose can pass through the cell membrane by means of a carrier protein, but glucose-6-phosphate cannot. The reaction catalyzed by hexokinase,

$$G + ATP \leftrightarrow G\text{-}6\text{-}P + ADP,$$

has a $\Delta G°$ of -4 kcal/mole. Therefore at equilibrium, if $[ATP] = [ADP]$

$$[G\text{-}6\text{-}P^{(2)}]_{eq} = 10^3[G^{(2)}]_{eq} \text{ while } [G\text{-}6\text{-}P^{(1)}] = 0$$

since hexokinase is not found outside the cell. If we define total glucose concentration, $[G_t] = [G] + [G\text{-}6\text{-}P]$, we find that at equilibrium

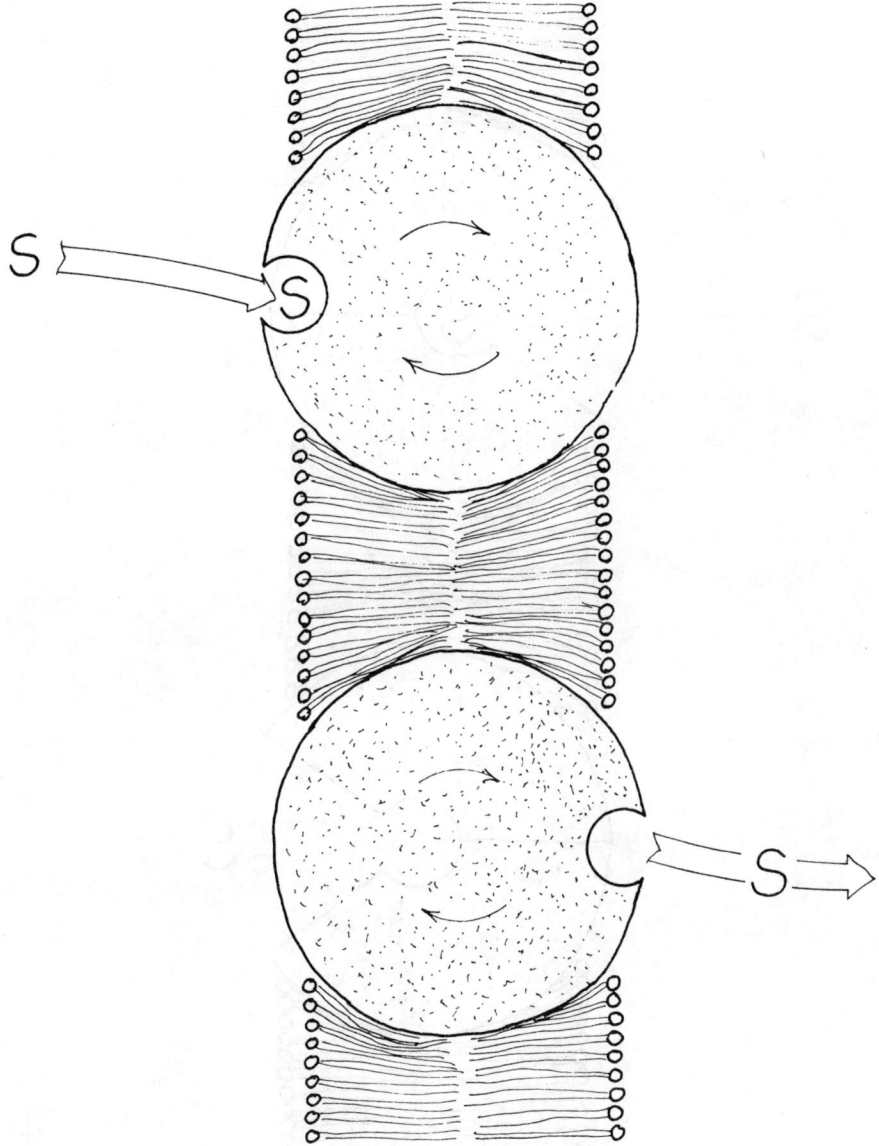

Fig. 5-3 Solute transport across a membrane by rotation of the transport protein in the membrane.

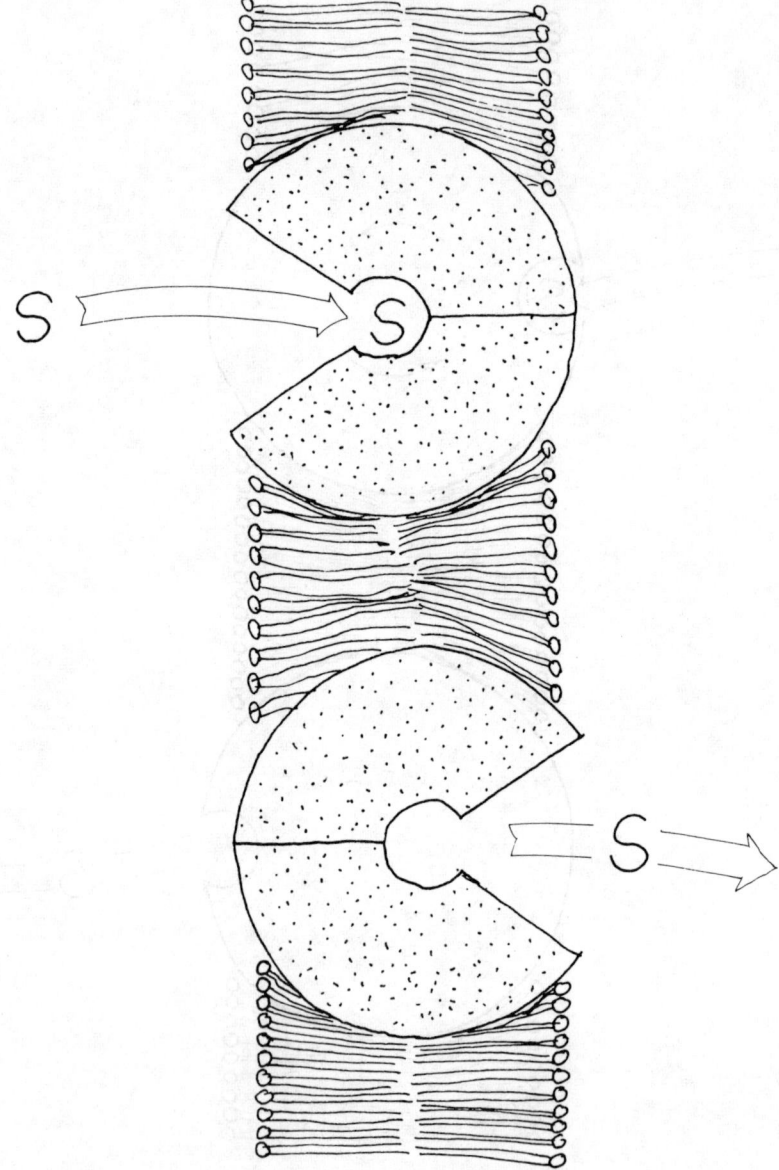

Fig. 5-4 Solute transport across a membrane by a reversible conformational transition in the transport protein.

$$[G_T^{(2)}] = 10^3 [G_T^{(1)}]$$

One could argue that this is not really pumping glucose against a concentration gradient since at equilibrium $[G^{(1)}] = [G^{(2)}]$. However, there can be no doubt that a large concentration difference in G-6-P is generated by this mechanism. One might think of this as the coupling of reactions in series. An alternative pattern of coupling, one which has more of the features of a pump, is parallel coupling.

Consider a transport protein which couples the cleavage of ATP to the translocation of solute S. Our two coupled reactions are

$$ATP + H_2O \rightarrow ADP + P_i \qquad \Delta G^0 = -7.3 \text{ kcal/mole}$$

$$S^{(1)} \rightarrow S^{(2)} \qquad \Delta G^0 = 0 \text{ kcal/mole}$$

to give the overall process

$$ATP + H_2O + S^{(1)} \rightarrow S^{(2)} + ADP + Pi \qquad \Delta G^0 = -7.3 \text{ kcal/mole}$$

When two such processes are coupled by an enzyme it means that the enzyme can promote or catalyze one of the processes only if the other also occurs. Under these conditions the enzyme can catalyze the movement of the system toward a minimum free energy or equilibrium of the summed process even when the individual reactions do not approach equilibrium.

For the summed process, the ratio of the equilibrium concentrations of $S^{(1)}$ and $S^{(2)}$ is a function of the concentrations of ATP, ADP and Pi. If these three reactants are at standard state, then

$$\frac{[S^{(1)}]_{eq}}{[S^{(2)}]_{eq}} = 1.5 \times 10^5$$

A large concentration gradient can be established by the investment of the energy of ATP hydrolysis.

A mechanism by which ATP dependent solute transport can be achieved can be easily envisioned by coupling a conformational change in the transport protein to phosphorylation and dephosphorylation. In this mechanism the unphosphorylated protein has a conformation which gives its binding site access to side 1 of the membrane. When S is bound to $C^{(1)}$ the carrier protein can be phosphorylated with ATP to give $SC^{(1)}P$ which undergoes a conforma-

$$C^{(1)} \xrightarrow{\quad S^{(1)} \quad} SC^{(1)} \xrightarrow{\quad \text{ATP} \quad \text{ADP} \quad} SC^{(1)}P$$

$$C^{(2)} \xleftarrow{\quad} C^{(2)}P \xleftarrow{\quad} SC^{(2)}P$$

$$\phantom{C^{(2)}} \xleftarrow{P_i} \phantom{C^{(2)}P} \xleftarrow{S^{(2)}}$$

tional change to $SC^{(2)}P$ with the binding site open to side 2 of the membrane. Dissociation of S permits the dephosphorylation of $C^{(2)}P$ which is thereby converted to $C^{(1)}$. This mechanism is similar to that presented in Fig. 5-4 for passive, carrier mediated diffusion except here the conformational changes in the carrier protein are driven, first from $C^{(1)}$ to $C^{(2)}$ by ATP dependent phosphorylation and then from $C^{(2)}$ to $C^{(1)}$ by dephosphorylation to yield inorganic phosphate. The total cycle from $C^{(1)}$ back to $C^{(1)}$ costs the net hydrolysis of one ATP to form ADP. The coupling of these processes requires that the phosphorylation of $C^{(1)}$ can occur only when S is bound and that the dephosphorylation of $C^{(2)}$ occurs only when S is not present on the protein. Uncoupling will result in the uncontrolled hydrolysis of ATP.

Such coupling of a chemical reaction to a transport process offers another possibility which must be discussed since it is of great importance in living systems. One should recall that enzymes merely increase the rate at which a system approaches equilibrium, i.e. dissipates excess free energy, without altering the final equilibrium state. That is, a reaction is catalyzed equally in both directions. Thus, if there is a sufficient difference between $[S^{(1)}]$ and $[S^{(2)}]$ these coupled reactions can be driven in reverse with the result that ATP will be synthesized from ADP and inorganic phosphate. For the system described this requires that $[S^{(2)}]/[S^{(1)}] > 1.9 \times 10^5$ if P_i, ATP and ADP are all at standard state. If they are at normal physiological concentrations, an even greater concentration difference would be required.

So far we have considered only uncharged solute molecules for which

$$\mu_i = \mu_i^0 + RT\ln c_i \tag{22}$$

is a reasonable approximation at low values of c_i, and

$$\Delta G = RT \ln \frac{c_i^{(2)}}{c_i^{(1)}} \tag{23}$$

for the transport of component i from compartment 1 to compartment 2. If we consider a solute with a net charge, Z, then if it or other charged solute molecules can be pumped from one side of the membrane to another, there is the possibility of generating a charge difference across the membrane and an electrical potential. Such a potential adds another term to the free energy change associated with the transport of a charged solute molecule or ion between compartments. It can be most easily dealt with by altering the equation for the chemical potential, μ_i, to give

$$\mu_i = \mu_i^0 + RT\ln c_i + ZFV \qquad (24)$$

where V is the electrical potential of the compartment and F is the Faraday constant. The free energy change associated with transporting such a solute from compartment 1 to compartment 2 is then given by

$$\Delta G = \mu_i^{(2)} - \mu_i^{(1)} = RT\ln \frac{c_i^{(1)}}{c_i^{(2)}} + ZF(V^{(2)} - V^{(1)}) \qquad (25)$$

Therefore, a large negative ΔG can be obtained if $c_i^{(1)} >> c_i^{(2)}$ and/or if $(V^{(2)} - V^{(1)})$ is large and has a sign opposite that of Z, the charge on i. Large free energy changes can be associated with ion translocation even in the absence of very large concentration differences. Given these considerations we are in a position to discuss the most fundamental processes of life, the generation of biologically useful energy from oxidation and the harvesting of the energy of the sun.

Oxidative Phosphorylation

Given an understanding of membrane structure, we are now in a position to understand the mechanism by which biological systems couple oxidation-reduction reactions to the formation of ATP. As we have already discussed, in the conversion of glucose to pyruvate two molecules of NAD^+ are reduced to NADH, and several more reducing equivalents are passed to NAD^+ during the oxidation of pyruvate to CO_2. In all, 10 moles of NADH are formed per mole of glucose oxidized.

The ultimate fate of the two extra electrons on NADH is to reduce an atom of oxygen to form water

$$NADH + H^+ + 1/2O_2 \rightarrow NAD^+ + H_2O; \qquad \Delta G^0 = -52.6 \text{ kcal/mole}$$

As one might have guessed, this transfer of electrons from NADH to oxygen is not carried out in a single step as written above. Instead, these reducing equivalents are passed sequentially to oxidizing agents with ever more positive redox potentials. The redox groups involved in these reactions include the following:

Flavin mononucleotide, FMN, is a two electron redox system. This is a prosthetic group of the enzyme NADH dehydrogenase. Reduction to $FMNH_2$ requires the uptake of two protons.

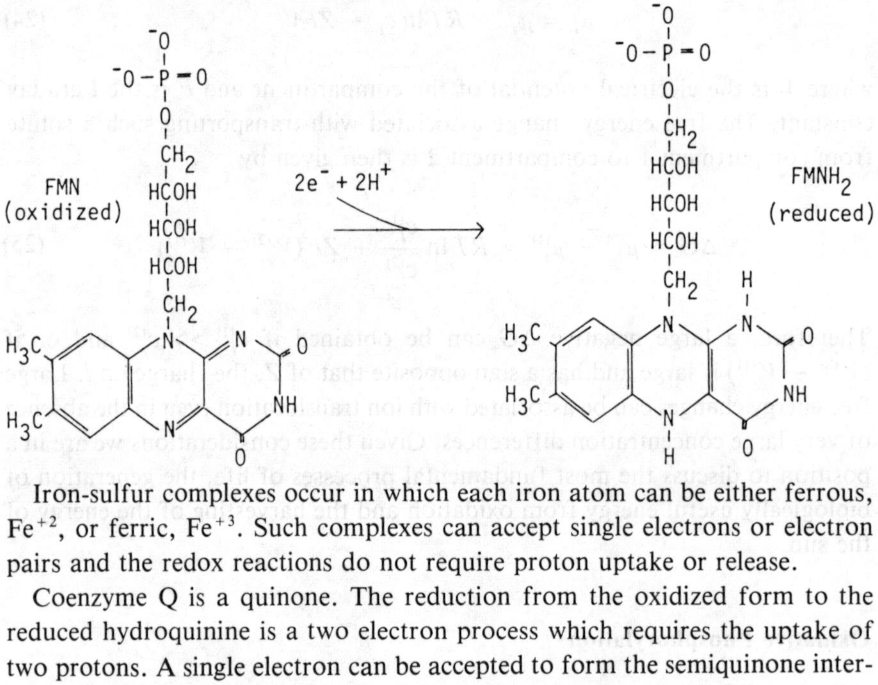

Iron-sulfur complexes occur in which each iron atom can be either ferrous, Fe^{+2}, or ferric, Fe^{+3}. Such complexes can accept single electrons or electron pairs and the redox reactions do not require proton uptake or release.

Coenzyme Q is a quinone. The reduction from the oxidized form to the reduced hydroquinine is a two electron process which requires the uptake of two protons. A single electron can be accepted to form the semiquinone intermediate which can be converted to the hydroquinone with another reducing equivalent.

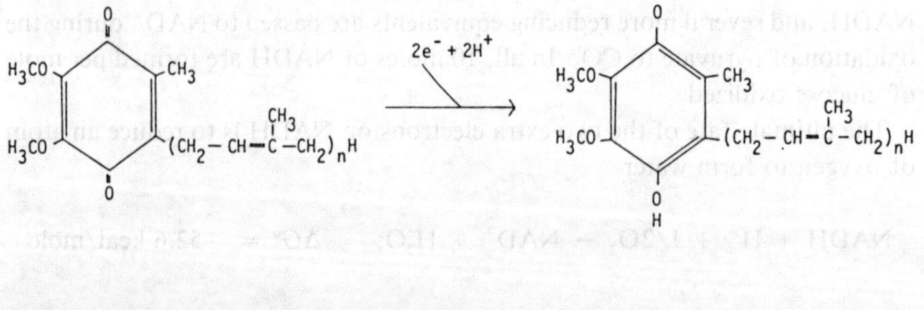

The cytochromes are heme proteins in which the heme iron can be either ferric or ferrous, i.e. each heme is a one electron redox system. There are three classes of cytochromes. Cytochromes *c* and *b* contain iron-protoporphyrin IX or heme. In cytochrome *c* the heme is covalently attached to the protein through thioether linkages between two cysteine residues and the vinyl groups of the heme. Cytochromes *a* contain heme A which differs from heme in having a hydrophobic side chain with three isoprenoid units instead of a vinyl group at position 2 and a formyl group instead of a methyl at position 8.

Heme (Fe-protoporphyrin IX)

Heme A

The standard reduction potentials for some of these systems are given in Table 5-I. For these potentials the standard state is 25° and 1M reactant concentrations with the exception of H^+ which is taken to be 10^{-7} M, i.e. pH = 7.

The reader will note that some of these systems require the uptake of protons for reduction and their release for oxidation while others do not. By interspersing proton-linked, two electron redox systems with single electron metal ion systems, one assures that protons are taken up and released rather than being passed down the electron transport chain. The enzymes of the electron transport chain are located in the inner membrane of the mitochondrion. By combining oxidation and reduction with the motion of electron carriers in the membrane and/or allosteric transitions in the structures of these enzymes a mechanism can be envisioned such that protons are always taken up for these reactions from the interior space or matrix of the mitochondrion and are always released on the exterior or cytoplasmic side of the membrane of this organelle. Thus these redox reactions serve to pump protons from the mitochondrial matrix into the cytoplasm producing an electrochemical gradient across the membrane. This gradient is then used essentially to drive an ATP dependent proton pump backward in order to produce ATP from ADP and inorganic phosphate.

TABLE 5-I

Redux Potentials

half cell	E_0' (Volts)*
$2H^+ + 2e^- \rightleftharpoons H_2$	-0.42
$NAD^+ + 2H^+ + 2e^- \rightleftharpoons NADH + H^+$	-0.32
$NADP^+ + 2H^+ + 2e^- \rightleftharpoons NADPH + H^+$	-0.32
$FAD + 2H^+ + 2e^- \rightleftharpoons FADH_2$	-0.18
Ubiquinone $+ 2H^+ + 2e \rightleftharpoons$ Ubiquinol	$+0.10$
2 cytochrome c $(Fe^{+3}) + 2e^- \rightleftharpoons$ 2 cytochrome c (Fe^{+2})	$+0.23$
$\frac{1}{2}O_2 + 2H^+ + 2e^- \rightleftharpoons H_2O$	$+0.82$

* For these potentials the standard state is taken to be 1 Molar reactant concentrations with the exception of the proton concentration which is taken as $10^{-7}M$, pH7. The hydrogen half cell would have a redox potential of 0.0 at 1 M protons, pH = 0.

The mitochondrion is a cellular organelle which possesses two membranes, see Fig. 5-5. The outer membrane appears to serve primarily as an envelope for the organelle. It contains numerous diffusion channels and acts as a diffusion barrier only to solutes of relatively high molecular weight, > 5000 daltons. The inner mitochondrial membrane surrounds the matrix of the organelle. Within the matrix or imbedded in the inner membrane are all of the enzymes for the oxidation of pyruvate, the electron-transport chain and the proton dependent ATPase by which ATP is synthesized by the energy of an electrochemical proton gradient. The inner membrane has a much greater surface area than the outer membrane, and therefore is folded back upon itself to form multiple layers or cristae. There are numerous mitochondria within a cell, their numbers being a reflection of the metabolic activity of the cell.

NADH in the interior of the mitochondrion interacts with the membrane bound enzyme, NADH reductase. The electrons from NADH along with its proton and one from solution are passed to the enzyme bound FMN to form $FMNH_2$.

$$NADH + H^+ + FMN \rightarrow NAD^+ + FMNH_2$$

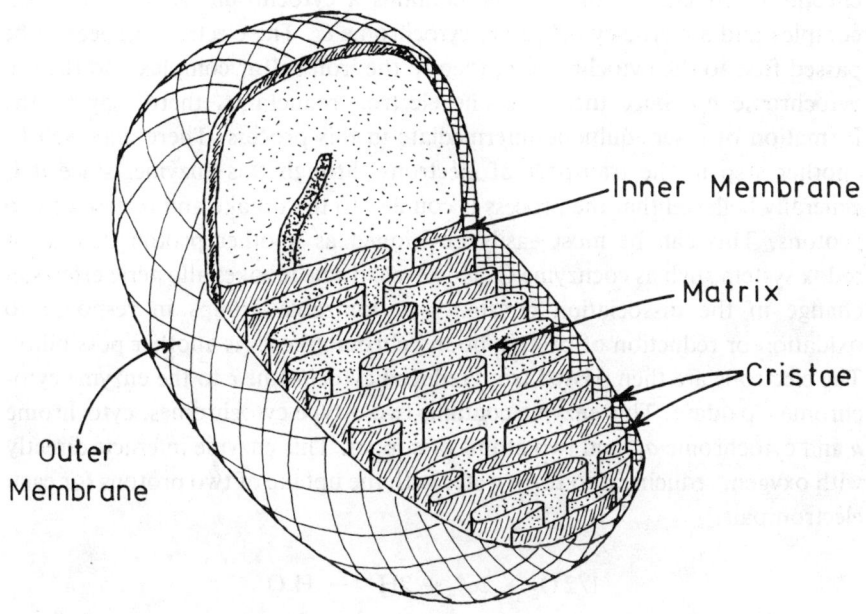

Fig. 5-5 A mitochondrion.

This enzyme also contains an iron-sulfur complex and it appears that the electrons are passed next to this complex. This requires the release of two protons.

$$FMNH_2 + 2(Fe^{3+} - S) \rightarrow FMN + 2(Fe^{2+} - S) + 2H^+$$

The electrons are next passed from NADH reductase to coenzyme Q to form the hydroquinone. This requires the uptake of two protons.

$$2(Fe^{2+}-S) + CoQ + 2H^+ \rightarrow 2(Fe^{3+}-S) + CoQ\ H_2$$

Reduced coenzyme Q now interacts with the membrane bound enzyme cytochrome c reductase. This enzyme contains a cytochrome b, an iron-sulfur complex and a c-type cytochrome, cytochrome c_1. The electrons appear to be passed first to the cytochrome b, then to the iron sulfur complex and then to cytochrome c_1. Since these are one electron reductants, there may be the formation of a semiquinone intermediate in this process. There may well be another step in the transport of electrons through this enzyme, since it is generally believed that the process is coupled to the uptake and release of two protons. This can be most easily envisioned as another proton dependent redox system such as coenzyme Q. However, if one invokes allosteric effects, a change in the dissociation constants of ionizable groups in response to oxidation or reduction of protein bound chromophores is another possibility. The electrons are then passed from reduced cytochrome c to the enzyme cytochrome c oxidase. This enzyme contains two a-type cytochromes, cytochrome a and cytochrome a_3, and two atoms of copper. This enzyme interacts directly with oxygen, reducing it to form water with the uptake of two protons for each electron pair.

$$1/2\ O_2 + 2e^- + 2H^+ \rightarrow H_2O$$

In Fig. 5-6 a possible scheme of the electron transport chain is presented indicating the directionality of proton uptake and release which could account for proton pumping and the establishment of an electrochemical proton gradient.

Proton translocation by cytochrome c reductase is shown although its mechanism is unknown. Nevertheless, a potential mechanism for the translocation of four protons for each electron pair is clear. The pH gradient and membrane potential produced across the inner mitochondrial membrane by the electron transport chain are of sufficient magnitude that the transport of 2 protons from the cytoplasm into the matrix will produce a large enough free energy change to drive the synthesis of one ATP from ADP and inorganic phosphate. It is generally agreed that the oxidation of one NADH produces three molecules of ATP, and this is consistent with the translocation of six protons for each electron pair that passes through the electron transport chain.

In the oxidation of pyruvate one electron pair is passed to the enzyme succinate dehydrogenase, where an enzyme-bound flavin adenine dinucleotide, FAD, is reduced to $FADH_2$. These electrons enter the electron transport

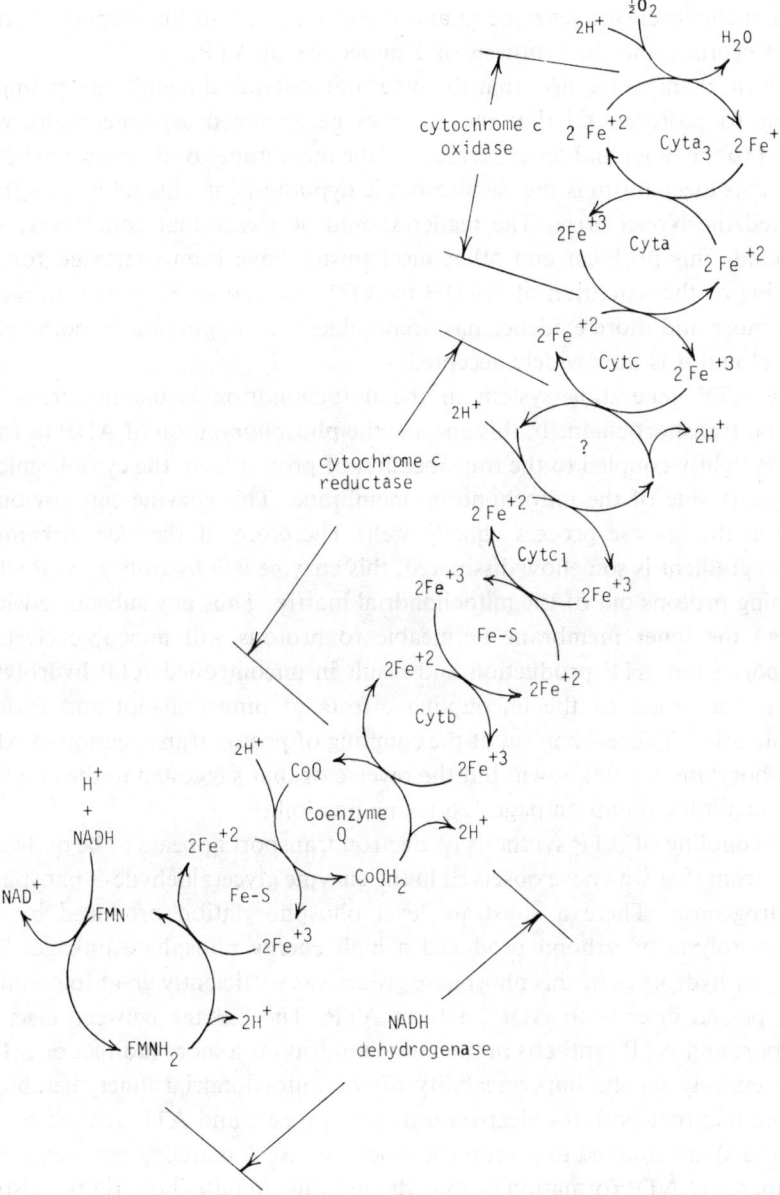

Fig. 5-6 The electron transport chain of the mitochondrion. The pathway followed by an electron pair from NADH to oxygen is shown. Also indicated are the points at which protons are removed from the matrix and released into the inner membrane space.

chain at the level of coenzyme Q and therefore result in the translocation of only 4 protons and the synthesis of 2 molecules of ATP.

This mechanism requires that the inner mitochondrial membrane be impermeable to protons and that the enzymes be arranged asymmetrically with respect to the inner and outer surfaces of the membrane. Both appear to be the case. This mechanism is the chemiosmotic hypothesis of Mitchel for which he received the Nobel prize. The reader should be aware that controversy still surrounds this problem and other mechanisms have been suggested for the coupling of the oxidation of NADH to ATP production. However, in recent years more and more evidence has accumulated to support the hypothesis of Mitchel and it is now widely accepted.

The ATP generating system in the mitochondrion is distinct from the electron transport chain. In this enzyme the phosphorylation of ADP to form ATP is tightly coupled to the translocation of proton from the cytoplasmic to the matrix side of the mitochondrial membrane. This enzyme can obviously catalyze the reverse process equally well. Therefore, if the electrochemical proton gradient is somehow dissipated, this enzyme will hydrolyze ATP while pumping protons out of the mitochondrial matrix. Thus any substance which renders the inner membrane permeable to protons will uncouple electron transport from ATP production and result in uncontrolled ATP hydrolysis. This is the origin of the uncoupling effects of dinitrophenol and various organic acids. The mechanism of the coupling of proton translocation to ADP phosphorylation is unknown, but the reverse of that suggested in the diagram of a membrane pump on page 126 is one possibility.

The coupling of ATP synthesis to electron transport appears to be quite different from that which we observed in the enzyme glyceraldehyde-3-phosphate dehydrogenase. There a substrate level phosphorylation produced by the phosphorolysis of a bond produced a high energy phosphate linkage. The energy of hydrolysis of this phosphate group was sufficiently great to permit it to be passed directly to ADP to form ATP. The linkage between electron transport and ATP synthesis in the mitochondrion is a more indirect one. It is based entirely on the impermeability of the mitochondrial inner membrane and the fact that both the electron transport process and ATP generation (or hydrolysis) are coupled to proton translocation. Such coupling between redox reactions and ATP formation is observed not only in mitochondria but also in another energy generating apparatus of living systems, chloroplasts. We are now in a position to study the most crucial of all processes to living systems, the process by which energy is obtained from the sun, photosynthesis.

Photosynthesis

The ultimate energy source for virtually all living systems on earth is the sun.* True, animals ingest other animals or plants to obtain food which can be metabolized to produce energy. However, if one follows any food chain from predator to a prey which may be the predator of another prey one eventually reaches herbivorous animals which derive their energy from plants which in turn obtain theirs from the sun. Therefore, the ecosystem of living organisms on earth constitutes a complex machine which is powered by solar energy and essentially participates in the dissipation of solar energy to the universe. The question with which we want to deal now is: How do plants convert solar energy into a biochemically useful form?

The majority of solar energy arrives at the earth in the form of electromagnetic radiation or photons. Most of the higher energy radiation, x-rays and far ultraviolet radiation, is filtered out by the atmosphere. This is fortunate since such radiation produces irreversible damage to the complex molecules of living systems. The radiation reaching the earth's surface is primarily in the visible as well as the near infrared and near ultraviolet regions of the spectrum. As is well known, light interacts with matter by exciting a transition between the quantized energy states of an atom or molecule. These energy states are defined by, in order of increasing energy differences, quantized translational, rotational, vibrational and electronic states. Because light energy is quantized, its interaction with matter has a significant probability only when it can induce a transition between states whose energy difference equals the energy of the photon.

$$E = h\nu \qquad (26)$$

There is also a requirement that the states differ in such a way that the dipole moment of the molecule is changed in the transition. For the vast majority of the photons reaching the earth's surface vibrational energy levels do not offer sufficiently large energy differences to permit interaction. On the other hand,

---

* The exception to this are organisms which are found at sites of geothermal activity on the ocean floor. At such "rift vents" there is heat and an abundance of compounds of reduced sulfur, particularly hydrogen sulfide. There are bacteria which can obtain metabolic energy from the oxidation of these compounds, and these serve as the foundation for the food chain in these ecosystems. Although not dependent on solar energy, these organisms are existing by utilizing energy which is being dissipated from the earth's core, energy left over from the formation of the solar system.

the differences between the energies of the electronic ground states and excited states of lowest energy are for most simple compounds or atoms too great to permit interaction and light absorption. Compounds in which available electronic states differ by sufficiently small energies are complexes of transition metals with their d-orbital systems, and molecules with extended π-orbitals, i.e. systems of conjugated double bonds. The principle pigments of photosynthetic organisms fall into this category. In green plants light is absorbed by two closely related pigments, chlorophyll *a* and chlorophyll *b*. The structures of these two compounds

differ only by the replacement of a methyl by a formyl group on one of the rings. These structures resemble heme, the prosthetic group of cytochromes *a*, *b* and *c*, and in biological systems are synthesized from protoporphyrin IX, the porphyrin of heme. Along with several other changes, they differ from heme by having a magnesium rather than an iron ion bound to the pyrrole nitrogens. With such an extended π electron system, these compounds interact strongly with visible radiation. The absorption spectrum of chlorophyll *a* is shown in Fig. 5-7.

Once a molecule is excited to a higher energy electronic state, it must dissipate the energy it has absorbed in order to return to its ground state. There are several pathways by which this can be accomplished. The molecule will quickly fall to the lowest vibrational ground state of the excited electronic state since vibrational modes are strongly coupled to the surroundings and the energy of these transitions is quickly dissipated to the surroundings as heat. If there are multiple electronic excited states whose energies overlap because of their multiple vibrational modes, then the molecule can cascade through these states always dissipating vibrational energy as heat. In compounds such as heme

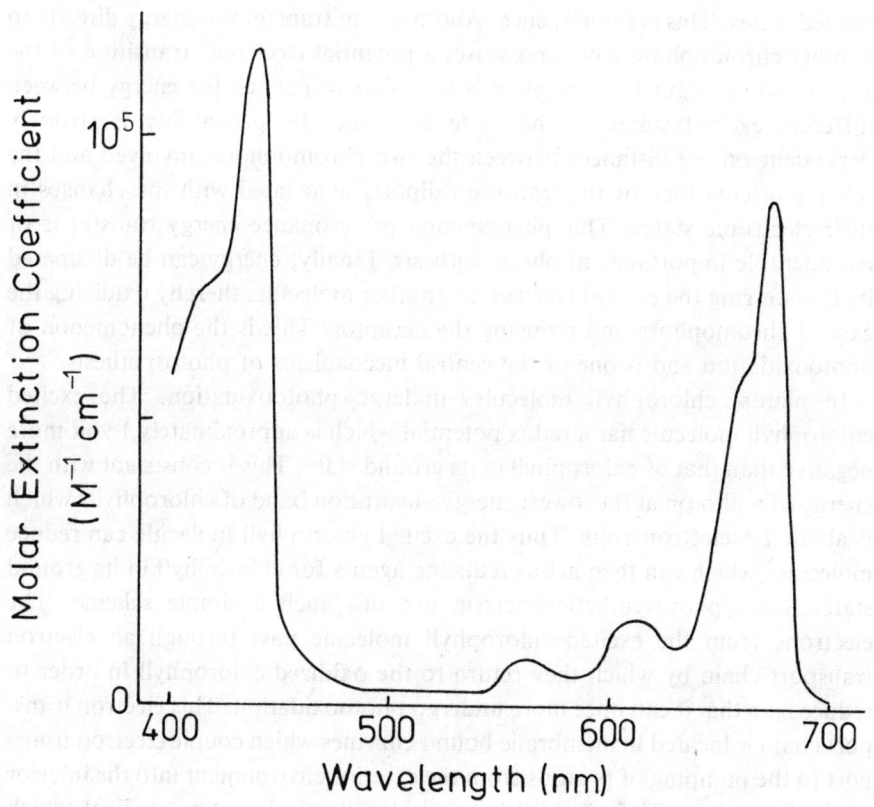

Fig. 5-7 The absorption spectrum of chlorophyll a.

there exists sufficient excited states of different energy that virtually all energy absorbed from light can be dissipated in this way. For most chromophores, however, there is a significant gap between the minimum energy of the lowest excited electronic state and the maximum energy of the electronic ground state. This gap can be bridged by several means. One of them is to release a photon of energy, $h\nu$, equal to the energy difference between the ground and excited states. This is fluorescence. Another is to transfer the energy directly to another chromophore which possesses a potential electronic transition of the appropriate energy. In principle this is similar to passing the energy between different excited states of the same molecule. Its probability is strongly dependent on the distances between the two chromophores involved and the relative orientations of the transition dipoles associated with the changes in their electronic states. This phenomenon of resonance energy transfer is of considerable importance in photosynthesis. Finally, energy can be dissipated by transferring the excited electron to another molecule, thereby oxidizing the excited chromophore and reducing the acceptor. This is the phenomenon of photooxidation and is one of the central mechanisms of photosynthesis.

In plants, chlorophyll molecules undergo photooxidation. The excited chlorophyll molecule has a redox potential which is approximately 1 volt more negative than that of chlorophyll in its ground state. This is consistent with the energy of a photon at the lowest energy absorption band of chlorophylls which is about 1.5 electron volts. Thus the excited chlorophyll molecule can reduce molecules which can then act as reducing agents for chlorophyll in its ground state. Some photosynthetic bacteria use just such a simple scheme. The electrons from the excited chlorophyll molecule pass through an electron transport chain by which they return to the oxidized chlorophyll in order to reduce it so that it can once more undergo photooxidation. This electron transport chain is located in membrane bound enzymes which couple electron transport to the pumping of protons from the exterior environment into the interior of the bacterium. This establishes an electrochemical proton gradient which can drive the synthesis of ATP in a manner analogous to that devised for oxidative phosphorylation in mitochondria.

Higher plants have developed a more complex system which appears to be more efficient. In these organisms the goal of photosynthesis is the reduction of atmospheric CO_2 by extracting reducing equivalents from water.

$$CO_2 + H_2O + h\nu \rightarrow (CH_2O) + O_2$$

where (CH_2O) indicates a carbon unit of a carbohydrate molecule such as glucose. This process can be carried out most simply with a reducing agent with a very negative redox potential and with ATP as an additional energy source. Plants obtain both from photooxidation and subsequent electron transport. The reducing agent which plants produce and use for CO_2 fixation is NADPH. This compound is identical to NADH with the exception of the attachment of a phosphate group to the number 2 carbon of the adenosyl ribose.

This compound

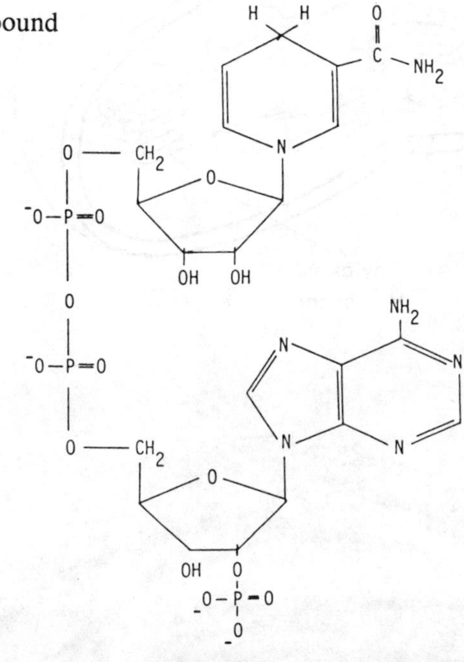

NADP

has virtually the same redox potential as NADH. The additional phosphate group serves merely for recognition. Normally NADPH is used as a reducing agent in biosynthetic reactions, not in oxidative phosphorylation for the production of ATP. Enzymes are specific for either NADPH or NADH. This permits the cell to keep the pool of biosynthetic reducing equivalents separate from that for ATP production and facilitates metabolic control. Therefore the intermediate reaction that plants achieve is

$$NADP^+ + H_2O + h\nu \rightarrow NADPH + H^+ + 1/2 O_2$$

However there is a problem. The difference between the redox potentials of $NADP^+$ and O_2 is about 1.1 volts. As we have already mentioned, the differ-

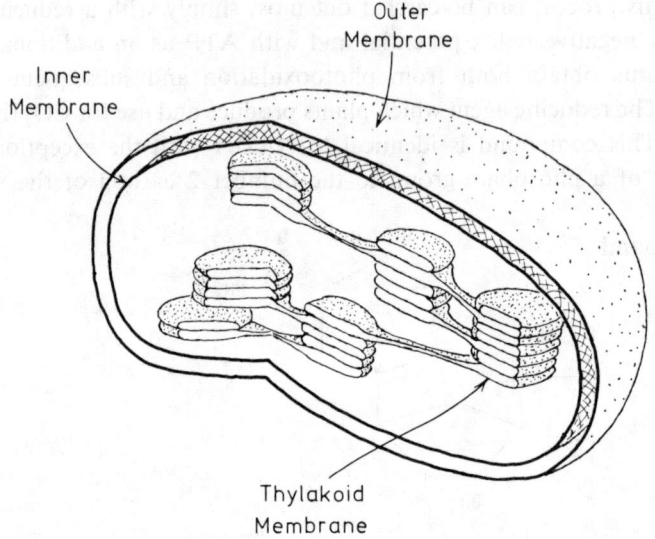

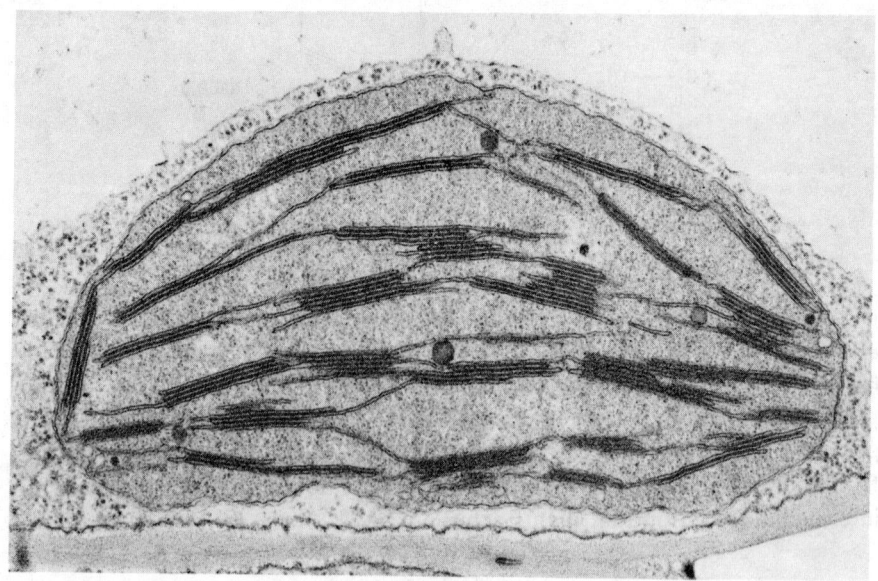

Fig. 5-8   The chloroplast. The electron micrograph was kindly supplied
by Professor L. Andrew Staehelin of the Department of Molecular,
Cellular and Development Biology, University of Colorado, Boulder. It
is a tobacco chloroplast at a magnification of 20,000×.

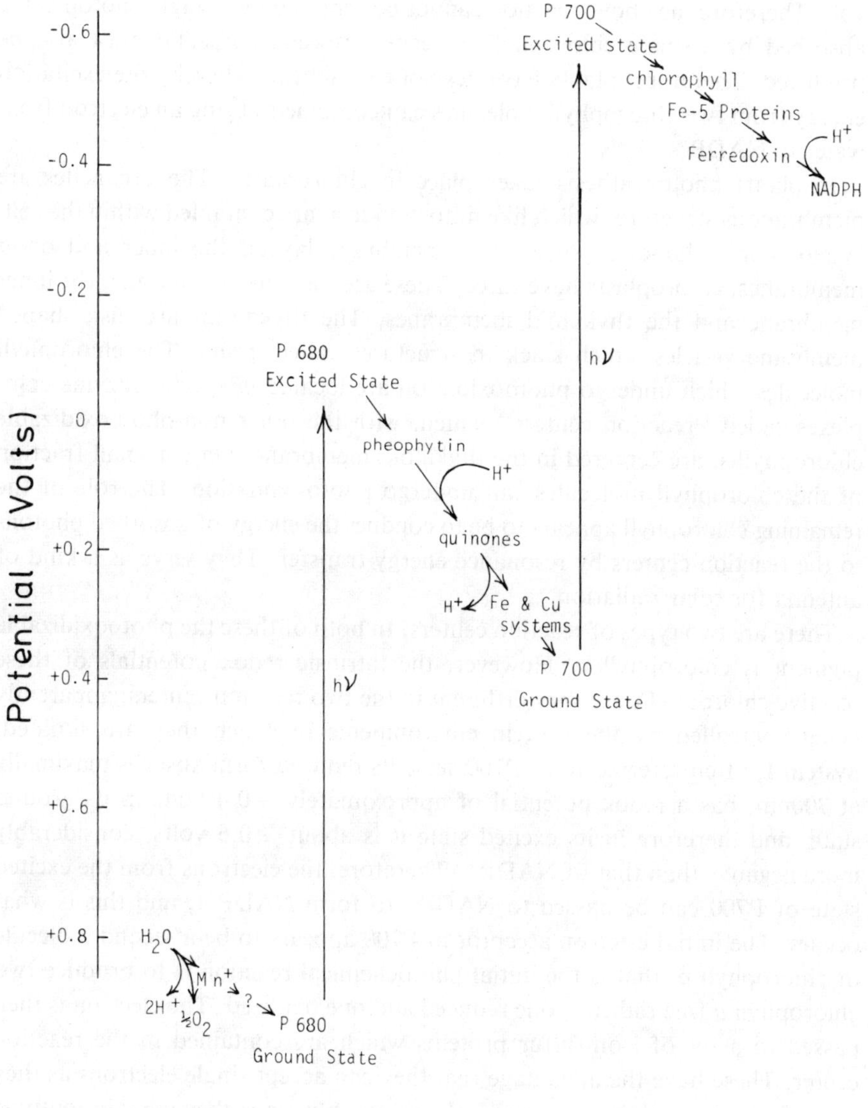

Fig. 5-9 The path of electrons in photosynthesis in higher plants. As in figure 5-6 the sites of uptake and release of protons are indicated.

ence in the redox potentials of chlorophyll in its ground and excited states is 1 volt. Therefore the above reaction cannot be achieved by a single photon being absorbed by a single chlorophyll molecule, especially if ATP must also be produced. Therefore, plants have developed a scheme whereby the excitation energy from two chlorophyll molecules can combine to bring an electron from water to $NADP^+$.

In plants photosynthesis takes place in chloroplasts. The organelles are membranous structure, which like mitochondria, are contained within the cell. Whereas mitochondria possess two membrane layers, the inner and outer membranes, chloroplasts have three. These are the outer membrane, the inner membrane and the thykaloid membranes. The thykaloids are disk shaped membrane vesicles which stack in structures called grana. The chlorophyll molecules which undergo photooxidation are located in special enzyme complexes called "reaction centers" which, with the other non-photooxidizable chlorophyllis, are centered in the thykaloid membrane. Only a small fraction of the chlorophyll molecules can undergo photo-oxidation. The role of the remaining chlorophyll appears to be to conduct the energy of absorbed photons to the reaction centers by resonance energy transfer. They serve as a kind of antenna for solar radiation.

There are two types of reaction centers. In both of these the photooxidizable pigment is chlorophyll *a*. However, the intrinsic redox potentials of these reactive chlorophylls are very different in the two reaction centers, apparently being controlled by the protein environments in which they are situated. System I, often referred to as P700 since its reduced form absorbs maximally at 700nm, has a redox potential of approximately $+0.4$ volts in its ground state, and therefore in its excited state it is about -0.6 volts, considerably more negative than that of $NADP^+$. Therefore, the electrons from the excited state of P700 can be passed to $NADP^+$ to form NADPH, and this is what occurs. The initial electron acceptor in P700 appears to be a second molecule of chlorophyll *a*, that is the initial photochemical reaction is to produce two chlorophyll *a* free radicals, one reduced and one oxidized. The electron is then passed to a set of iron-sulfur proteins which are contained in the reaction center. These have the advantage that they can accept single electrons as they come from the initial photochemical process, but since they contain multiple irons in their iron-sulfur centers they can store more than one reducing equivalent. The electrons are then passed to a soluble iron-sulfur protein, ferridoxin, which is found in the stroma, the space surrounded by the inner chloroplast membrane external to the thylakoids. Two electrons are passed from reduced ferredoxin to free $NADP^+$ by means of the enzyme, ferredoxin-

NADP oxidoreductase, which itself contains a flavin group. In passing an electron pair to this flavin protons are taken up from the stroma.

The second type of reaction center is called system II or P680 for the absorption maximum of its reduced form. This center furnishes electrons to reduce P700 so that the latter can continue to undergo photooxidation. System II obtains its electrons from water following photooxidation. Since the E_0 of the O_2–H_2O redox system is $+0.82$ volts the ground state of this reaction center must have a E_0 which is more positive than this. How a chlorophyll molecule can be made to be such a strong oxidant is not understood, but it is a property conferred by the protein in which the chlorophyll is bound. In P680 the first moiety to be reduced by the excited chlorophyll molecule appears to be pheophytin. This substance has the same structure as chlorophyll but lacks a magnesium ion. It accepts a single electron at a time to form a free radical ion. The electron is then passed through a series of plastoquinone groups.

$$H_3C \quad \overset{O}{\underset{O}{\bigcirc}} \quad H_3C \quad (CH_2-CH=\overset{CH_3}{\underset{}{C}}-CH_2)_n H$$

These have a structure similar to that of ubiquinone and can accept electrons one at a time by forming a semiquinone intermediate. They are proton dependent redox centers and protons are taken up when they accept electrons from pheophytin. From the plastoquinones electrons are passed through a set of transition metal centers which include heme proteins, iron-sulfur centers and copper containing proteins and ultimately reach the photo oxidized P700 system. When electrons pass from the quinone to these metal ions, protons are released. The mechanism by which electrons are transferred from water to P680 is not understood. An enzyme system which contains several manganese ions appears to be involved. The oxidation of water results in the release of protons.

As already mentioned P680 and P700 are contained in the thylakoid membrane. They are asymmetrically oriented so that protons are always released into interior space of the thylakoid and protons are always taken up from the stroma. This produces our familiar electrochemical proton gradient which in turn drives ATP production. Thus the chloroplast produces both NADPH and ATP and releases oxygen upon exposure to light.

Given a supply of NADPH and ATP, the fixation and reduction of CO_2 which takes place in the stroma is carried out by enzymatic catalysis as outlined below.

$$
\begin{array}{c}
CO_2 + \begin{array}{l} CH_2O \;\text{(P)} \\ | \\ C=O \\ | \\ HC-OH \\ | \\ HC-OH \\ | \\ CH_2O \;\text{(P)} \end{array}
\longrightarrow
\begin{array}{l} O \\ \backslash C-C-OH \\ -O/ \;\;| \\ CH_2O\text{(P)} \end{array}
\begin{array}{l} CH_2O \;\text{(P)} \\ | \\ C=O + H_2O \\ | \\ H-C-OH \\ | \\ CH_2O \;\text{(P)} \end{array}
\longrightarrow
2H-\begin{array}{l} CH_2O \;\text{(P)} \\ | \\ C-OH \\ | \\ COO^- \end{array}
\end{array}
$$

ribulose-1,5-diphosphate   transient intermediate   3-phosphoglycerate

$$
\begin{array}{l} CHO \\ | \\ HCOH \\ | \\ CH_2O \;\text{(P)} \end{array}
\xleftarrow[\;NADP^+ + P_i \qquad NADPH\;]{}
\begin{array}{l} CH_2O \;\text{(P)} \\ | \\ HC-OH \\ | \\ C-O \;\text{(P)} \\ \parallel \\ O \end{array}
$$

glyceraldehyde-3-phosphate   1,3-diphosphoglycerate

(ATP → ADP)

1,5-diphosphate of ribulose, a five carbon sugar, reacts with CO_2 to ultimately yield two molecules of 3-phosphoglycerate. The enzyme which catalyzes this reaction is ribulose diphosphate carboxylase. The equilibrium for this reaction lies strongly in the direction written with $\Delta G^{0\prime} = -12.4$ kcal/mole. Each 3-phosphoglycerate formed from ribulose diphosphate is then converted to glyceraldehyde-3-phosphate with the conversion of one ATP to ADP and the oxidation of one NADPH by essentially the reversal of this portion of the glycolytic cycle with the exception that the glyceraldehyde-3-phosphate dehydrogenase used in photosynthesis is specific for NADPH and will not accept NADH as a substrate. In glyceraldehyde-3-phosphate, all of the carbon atoms are at the oxidation state of carbohydrate. For every three molecules of CO_2 fixed in this manner, one uses three molecules of ribulose 1,5-diphosphate and obtains

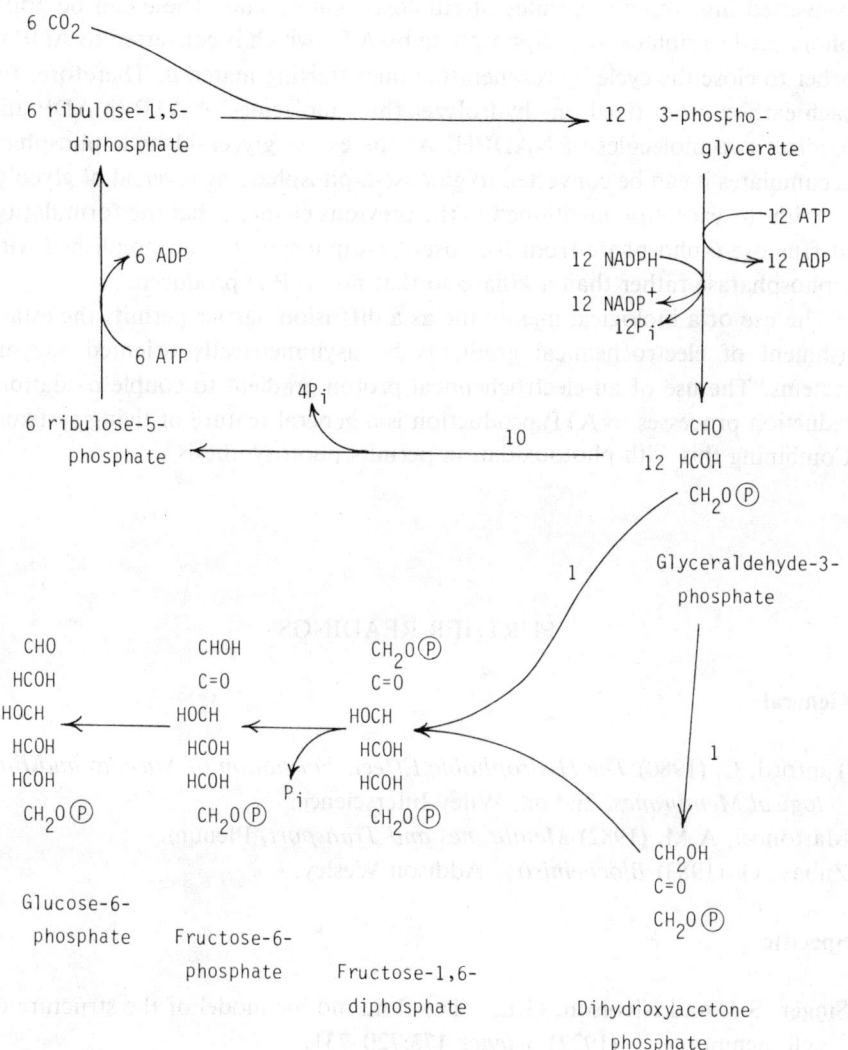

Fig. 5-10 The synthesis of glucose from CO_2 by oxidation of NADPH and hydrolysis of ATP.

6 molecules of glyceraldehyde-3-phosphate. By a series of isomerization and condensation reactions five molecules of glyceraldehyde-3-phosphate can be converted into three molecules of ribulose-5-phosphate. These can be phosphorylated to ribulose-1,5-diphosphate by ATP which is converted to ADP in order to close the cycle by regenerating ones starting material. Therefore, for each carbon atom fixed one hydrolyzes three molecules of ATP to ADP and oxidizes two molecules of NADPH. As the excess glyceraldehyde-phosphate accumulates it can be converted to glucose-6-phosphate by reversal of glycolysis with the exception mentioned in the previous chapter, that the formulation of fructose-6-phosphate from fructose-1, 6-diphosphate is accomplished with a phosphatase rather than a kinase so that no ATP is produced.

The use of a biological membrane as a diffusion barrier permits the establishment of electrochemical gradients by asymmetrically oriented enzyme systems. The use of an electrochemical proton gradient to couple oxidation-reduction processes to ATP production is a general feature of living systems. Combining this with photooxidation permits photosynthesis.

FURTHER READINGS

General

Tanford, C. (1980) *The Hydrophobic Effect: Formation of Micelles and Biological Membranes*, 2nd ed. Wiley-Interscience.

Martonosi, A.M. (1982) *Membranes and Transport*, Plenum.

Zubay, G. (1983) *Biochemistry*, Addison-Wesley.

Specific

Singer, S.J. and Nicolson, G.L. "The fluid mosaic model of the structure of cell membranes", (1972) *Science* **175**:720–731.

Mitchell, P. "Keilin's respiratory chain concept and its chemiosmotic consequences", (1979) *Science* **206**:1148–1159.

Hinkle, P.C. and McCarty, R.E. "How cells make ATP", (1978) *Sci. Amer.* **238**:104–123.

Fillingham, R.H. "The proton-translocating pumps of oxidative phosphorylation", (1980) *Annu. Rev. Biochem.* **49**:1079.

Wilson, D.F., Erecinska, M. and Dutton, P.L. "Thermodynamic relationships in mitochondrial oxidative phosphorylation", (1974) *Annu. Rev. Biophys. Bioeng.* **3**:203.

Arron, M. "Energy transduction in Chloroplasts", (1977) *Annu. Rev. Biochem.* **46**:143–155.

Edmond, J.M. and Von Damm, K. "Hot springs on the ocean floor", (1983) *Sci. Amer.* **248(4)**:78.

CHAPTER 6

POLYMERIC MOLECULES OF BIOLOGICAL SYSTEMS

> *"... the development and functioning of an organism consist essentially of an integrated system of chemical reactions controlled in some manner by genes."*
>
> Beadle and Tatum, 1941

Polymeric molecules of biological systems

There are many properties which are characteristic of living systems. However, one which appears to be unique to living systems, at least as we know them, is the occurrence of polymeric molecules which contain dissimilar units the sequence of which in the polymer constitutes readable information. It is these structures which we shall discuss in this chapter.

The best known and most important example of such a molecule is deoxyribonucleic acid, DNA. This is the stuff of genes, the substance of heredity, the material which can contain all the information for the construction of an organism, such as man, from a single cell. As already discussed in Chapter one, in examining life on earth it is hard not to define living things as those objects which in the process of self duplication duplicate a molecule of DNA or its close relative ribonucleic acid, RNA.

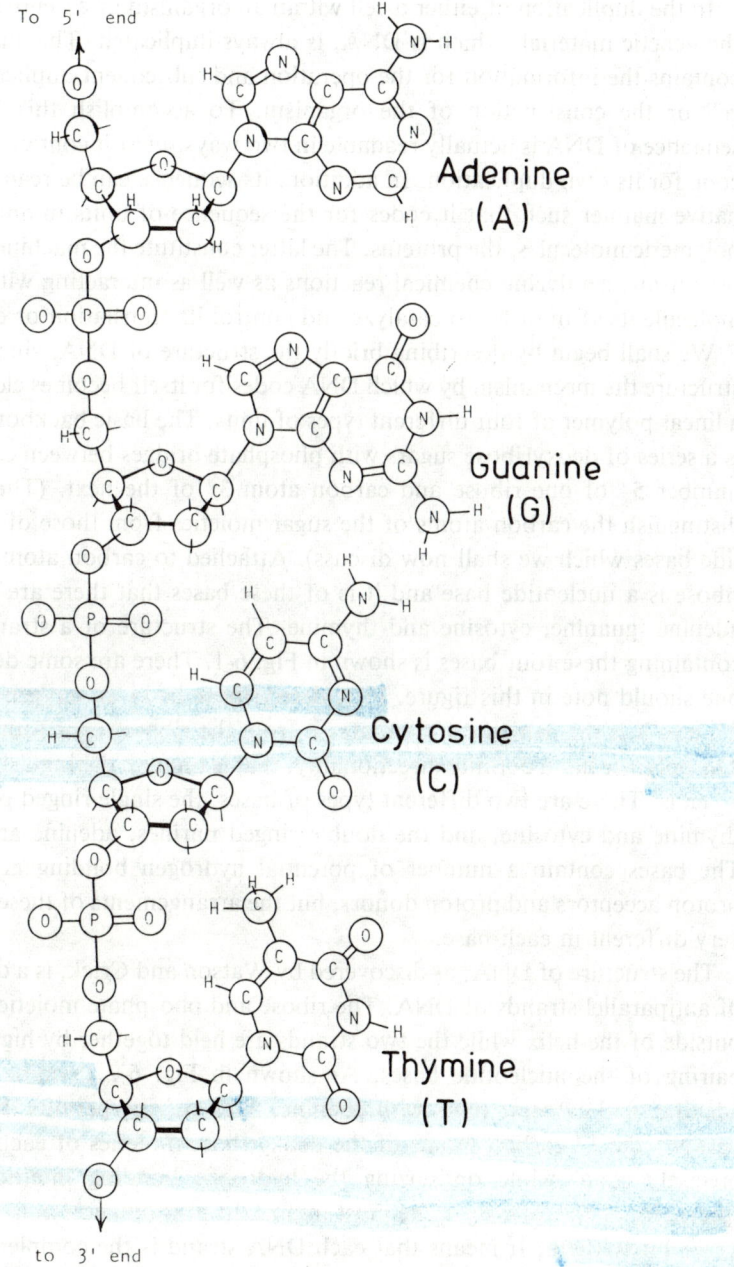

To 5' end

Adenine
(A)

Guanine
(G)

Cytosine
(C)

Thymine
(T)

to 3' end

Fig. 6-1 A single strand of DNA showing the 4 nucleotide bases.

In the duplication of either a cell within an organism or an entire organism, the genetic material, which is DNA, is always duplicated. This material then contains the information for the operation and subsequent duplication of the cell or the construction of the organism. To accomplish this the subunit sequence of DNA is actually readable in two ways. First it contains a readable code for its own duplication. In addition, its sequence can be read in an alternative manner such that it codes for the sequence of units in another set of polymeric molecules, the proteins. The latter constitute the machinery of living organisms, catalyzing chemical reactions as well as interacting with the DNA molecule itself in order to catalyze and control its translation or expression.

We shall begin by describing briefly the structure of DNA, since from this structure the mechanism by which DNA codes for itself becomes clear. DNA is a linear polymer of four different types of units. The basic backbone structure is a series of deoxyribose sugars with phosphate bridges between carbon atom number 5 ' of one ribose and carbon atom 3 ' of the next. (The "primes" distinguish the carbon atoms of the sugar moieties from those of the nucleotide bases which we shall now discuss). Attached to carbon atom 1 ' of each ribose is a nucleotide base and it is of these bases that there are four types, adenine, guanine, cytosine and thymine. The structure of a strand of DNA containing these four bases is shown in Fig. 6-1. There are some details which one should note in this figure. First deoxyribose lacks an oxygen at position 2 ', and it is this which differentiates it from the ribose sugar found in RNA. The polymer has a definite directionality, that is one can distinguish the 3 ' and 5 ' ends. There are two different types of bases, the single ringed pyrimidines, thymine and cytosine, and the double ringed purines, adenine and guanine. The bases contain a number of potential hydrogen bonding groups, both proton acceptors and proton donors, but the arrangements of these groups are very different in each base.

The structure of DNA, as discovered by Watson and Crick, is a double helix of antiparallel strands of DNA. The ribose and phosphate moieties lie on the outside of the helix while the two strands are held together by highly specific pairing of the nucleotide bases. As shown in Fig. 6-2 thymine pairs with adenine, and cytosine pairs with guanine. Such pairing permits the distances between the 1 ' carbon atoms of the deoxyribose moieties of each pair to be precisely equal while optimizing the hydrogen bonding interactions. The specificity of this pairing is the most important phenomenon in living systems as we know them. It means that each DNA strand is the complement of the other strand in the molecule. Each strand can act as a template for the duplication of its complementary strand. When each does this, the result is in effect

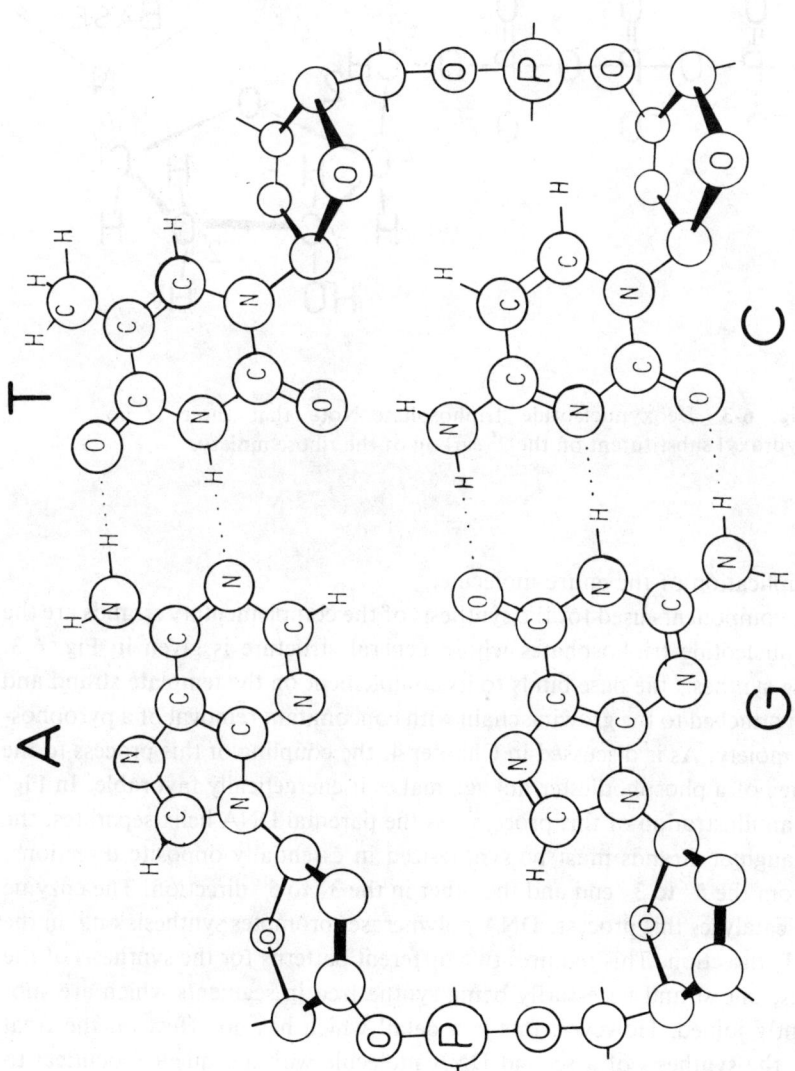

Fig. 6-2 Base pairing. The two pairs are adenine with thymine (top) and guanine with cytosine.

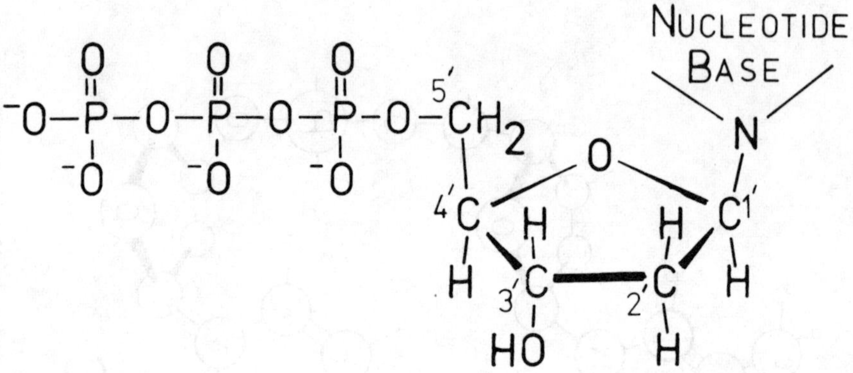

Fig. 6-3　Deoxynucleotide triphosphate. Note that there is no hydroxyl substitutent on the $2'$ carbon of the ribose moiety.

the duplication of the entire molecule.

The components used for the synthesis of the complementary strands are the deoxynucleotide triphosphates whose general structure is given in Fig. 6-3. During synthesis the base binds to its complement on the template strand and is then attached to the growing chain with concomitant removal of a pyrophosphate moiety. As is discussed in Chapter 4, the coupling of this process to the cleavage of a phosphodiester linkage makes it energetically favorable. In Fig. 6-4 is an illustration of this process. As the parental DNA helix separates, the two daughter strands must be synthesized in essentially opposite directions, one from the 5 ′ to 3 ′ end and the other in the 3 ′ to 5 ′ direction. The enzyme which catalyzes this process, DNA polymerase, promotes synthesis only in the 5 ′ to 3 ′ direction. This requires two different patterns for the synthesis of the strands, one strand necessarily being synthesized in segments which are subsequently joined. However, this is a detail which has no effect on the final result, the synthesis of a second DNA molecule with a sequence identical to that of the first. This process of DNA duplication occurs prior to each occurrence of cell division, and each of the cells produced in division receives one copy of the genome.

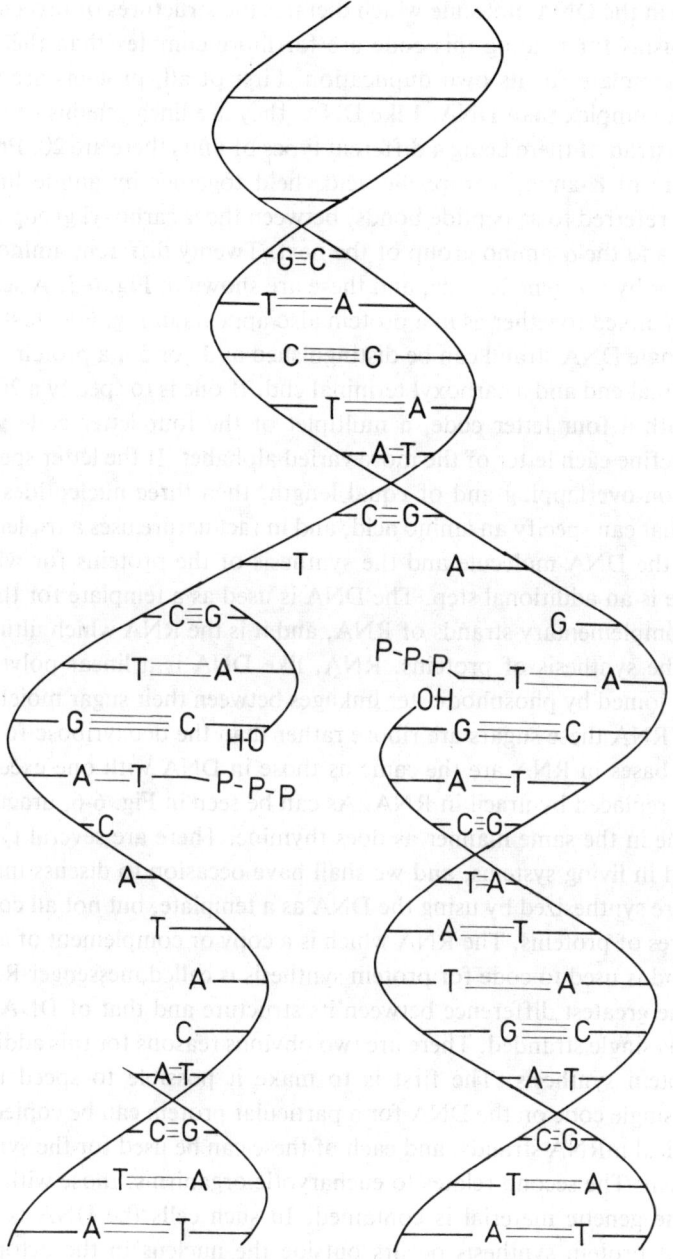

Fig. 6-4 Replication of DNA.

The code in the DNA molecule which dictates the structures of proteins and the mechanisms for reading this code are far more complex than the use of DNA as a template for its own duplication. First of all, proteins are intrinsically more complex than DNA. Like DNA, they are linear chains of similar units, but instead of there being 4 different types of units there are 20. Proteins are polymers of α-amino carboxylic acids held together by amide linkages (commonly referred to as peptide bonds) between the α carboxyl group of one amino acid and the α-amino group of the next. Twenty different amino acids are coded for by the genetic code, and these are shown in Fig. 6-5. A series of amino acids linked together as in a protein also appears in Fig. 6-5. Just as the ends of a single DNA strand can be distinguished as 3 ' or 5 ', a protein has an amino terminal end and a carboxyl terminal end. If one is to specify a 20 letter alphabet with a four letter code, a multiplet of the four letter code will be needed to define each letter of the more varied alphabet. If the letter specifiers are to be non-overlapping and of equal length, then three nucleotides is the minimum that can specify an amino acid, and in fact nature uses a triplet code.

Between the DNA molecule and the synthesis of the proteins for which it codes, there is an additional step. The DNA is used as a template for the synthesis of complementary strands of RNA, and it is the RNA which ultimately codes for the synthesis of proteins. RNA, like DNA is a linear polymer of nucleotides joined by phosphodiester linkages between their sugar moieties. In the case of RNA these sugars are ribose rather than the deoxyribose found in DNA. The bases in RNA are the same as those in DNA with one exception. Thymine is replaced by uracil in RNA. As can be seen in Fig. 6-6, uracil pairs with adenine in the same manner as does thymine. There are several types of RNA found in living systems, and we shall have occasion to discuss many of them. All are synthesized by using the DNA as a template, but not all code for the structures of proteins. The RNA which is a copy or complement of a DNA sequence and is used to code for protein synthesis is called messenger RNA or mRNA. The greatest difference between its structure and that of DNA is the fact that it is single stranded. There are two obvious reasons for this additional step in protein synthesis. The first is to make it possible to speed up the process. A single code on the DNA for a particular protein can be copied onto many identical mRNA strands, and each of these can be used for the synthesis of the protein. The second relates to eucharyotic organisms, those with nuclei in which the genetic material is contained. In such cells the DNA is in the nucleus, but protein synthesis occurs outside the nucleus in the cytoplasm. This can only be done by interposing a messenger between the DNA and the process of protein synthesis.

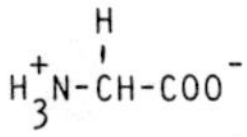

$$\overset{H}{\underset{H_3\overset{+}{N}-CH-COO^-}{|}}$$

Glycine (Gly)

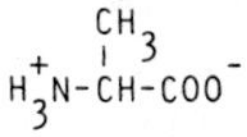

$$\overset{CH_3}{\underset{H_3\overset{+}{N}-CH-COO^-}{|}}$$

Alanine (Ala)

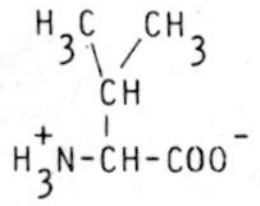

$$\overset{H_3C\diagdown\diagup CH_3}{\underset{H_3\overset{+}{N}-CH-COO^-}{\overset{|}{CH}}}$$

Valine (Val)

$$\overset{H_3C\diagdown\diagup CH_3}{\underset{H_3\overset{+}{N}-CH-COO^-}{\overset{|}{\underset{CH_2}{CH}}}}$$

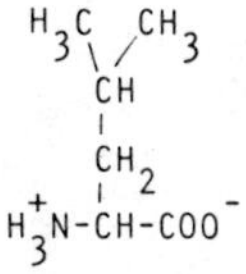

Leucine (Leu)

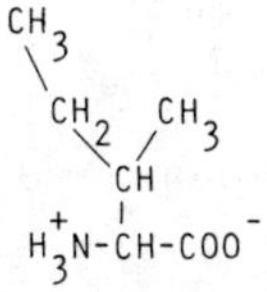

Isoleucine (Ile)

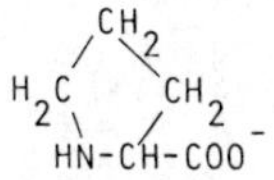

Proline (Pro)

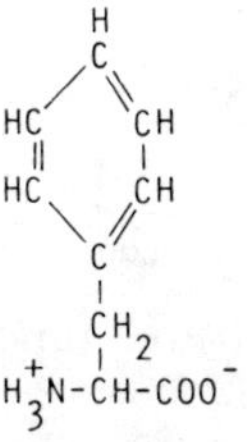

Phenylalanine (Phe)

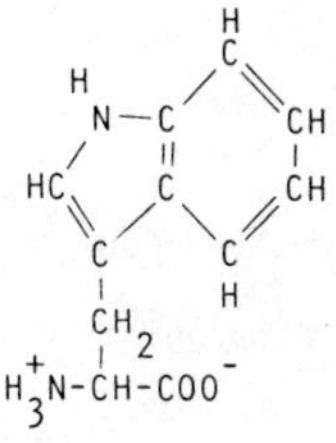

Tryptophan (Trp)

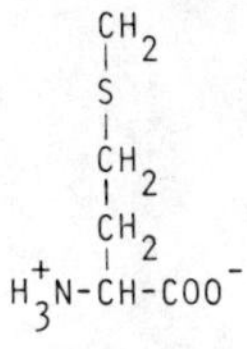

Methionine (Met)

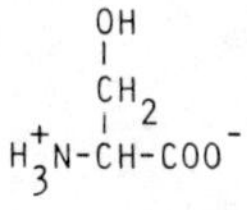

Serine (Ser)

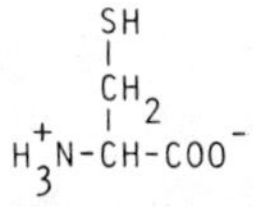

Cysteine (Cys)

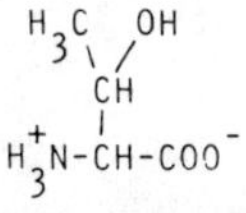

Threonine (Thr)

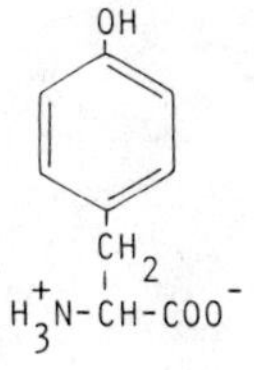

Tyrosine (Tyr)

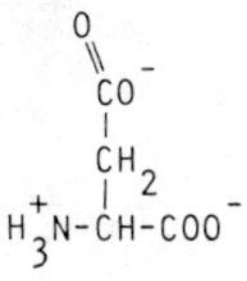

Aspartate (Asp)

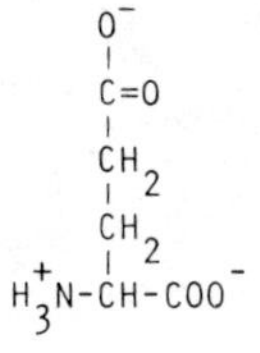

Glutamate (Glu)

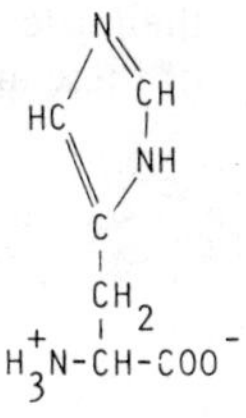

Histidine (His)

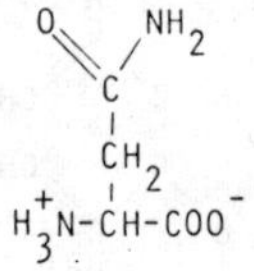

Asparagine (Asn) 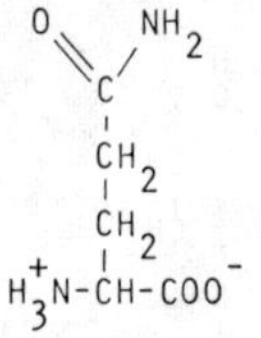 Glutamine (Glu) 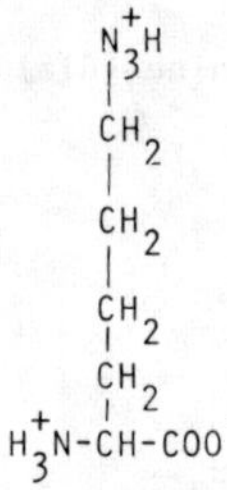 Lysine (Lys) 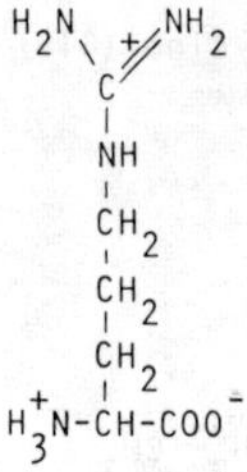 Arginine (Arg)

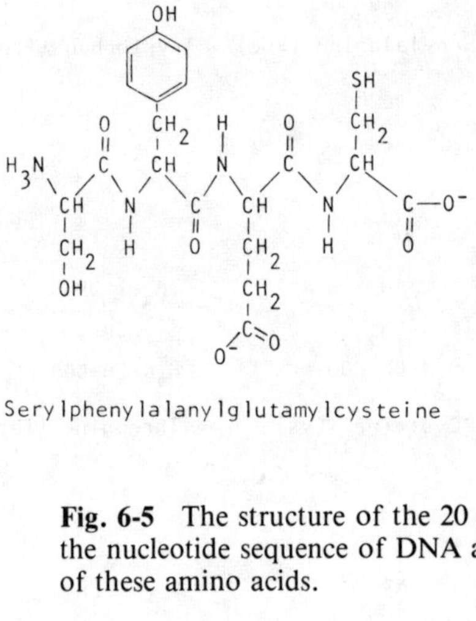

Serylphenylalanylglutamylcysteine

Fig. 6-5 The structure of the 20 amino acids which can be coded for by the nucleotide sequence of DNA and the structure of a peptide of some of these amino acids.

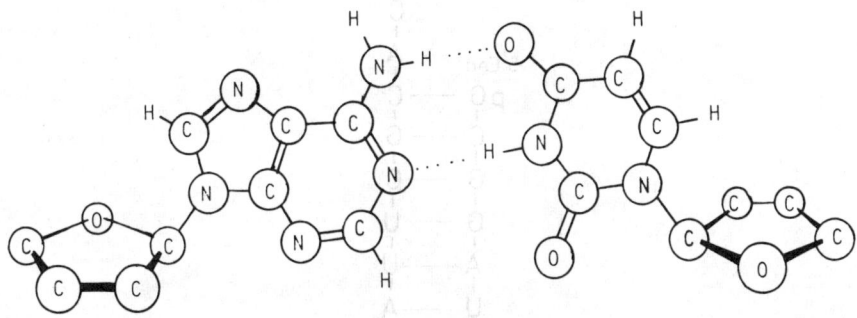

Fig. 6-6 The pairing of thymine with uracil.

Given the strand of mRNA, it is necessary to translate the sequence of nucleotide bases into a sequence of amino acids. The keys to this code are a set of small RNA molecules called transfer RNA or tRNA. Cells contain numerous different types of tRNA, each of which has two specificities. Each contains an anticodon triplet which is complementary to a triplet codon on the mRNA; and each can have attached to it, by highly specific enzymatic catalysis, one particular amino acid. In Fig. 6-7 is shown the primary and secondary structure of one tRNA, the phenylalanyl tRNA or tRNA^{phe}, of the bacterium *E. coli*. As can be seen, the sequence of nucleotides in tRNA is such as to allow base pairing between adjacent sequence elements with the resultant formation of "hairpin" like structures. These structures then fold up into a precisely defined three dimensional structure as shown in Fig. 6-8.

The number II hairpin loop, also known as the anticodon loop, has at its end the nucleotide triplet which pairs with the specific codon triplet on mRNA. This tRNA can have only the phenylalanine residue attached to the ribose of its 3 ′ terminal adenine group. The enzyme that does this recognizes both the specific amino acid and the specific tRNA and catalyzes only the correct attachment. At the same time its specific anticodon sequence, GAA (read in

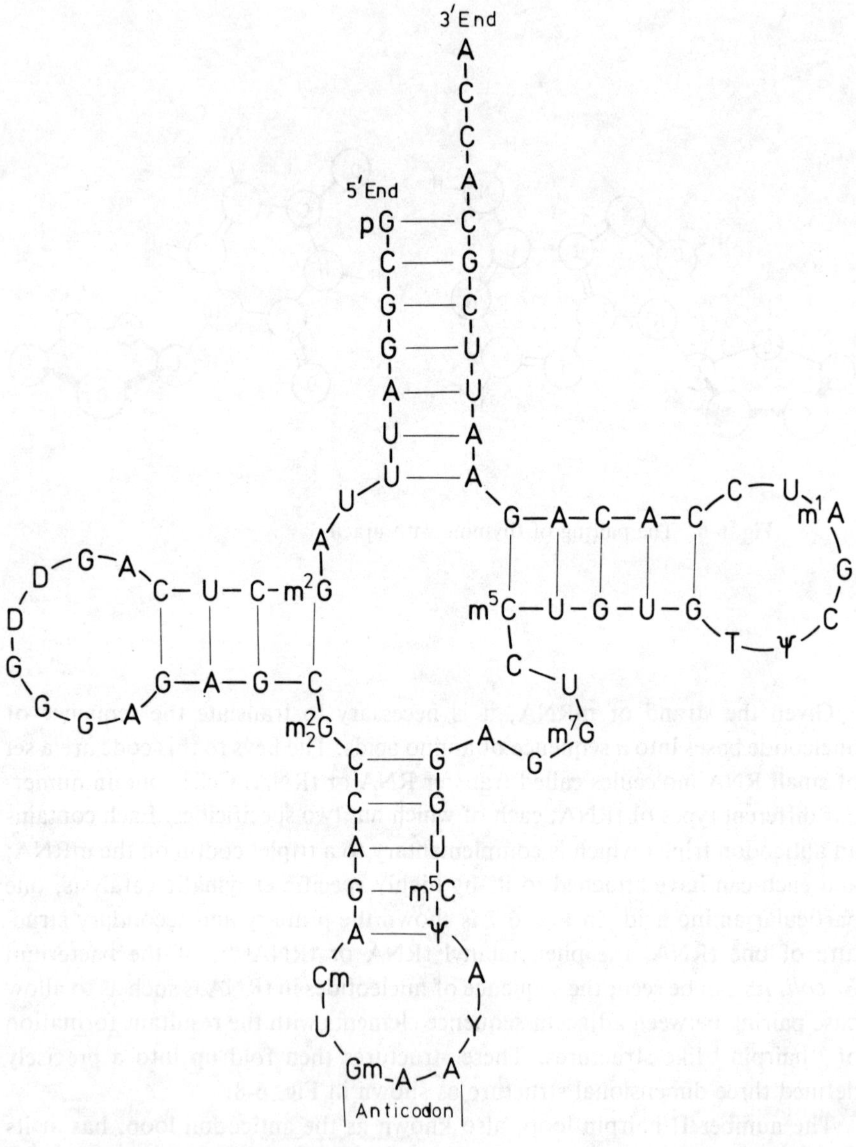

Fig. 6-7   Transfer ribonucleic acid, tRNA, Here one sees the primary and secondary structure of phenylalanyl tRNA. Th m refers to a methylated derivative of a nucleotide base. Unusual bases are denoted as follows: ψ-pseudouridine, I - inosine, D - dihydrouridine, T - ribothyimidine.

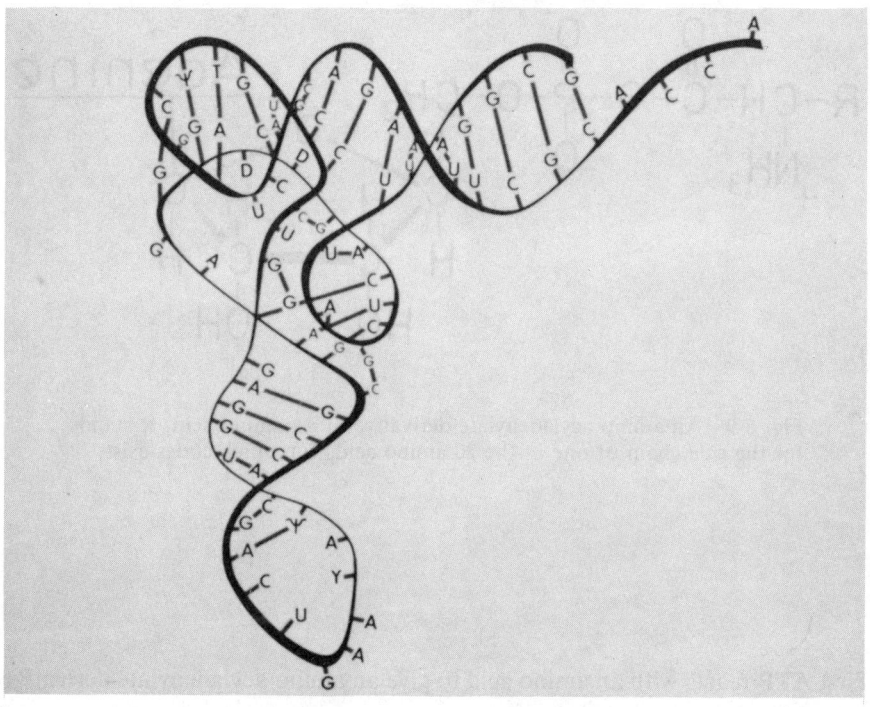

Fig. 6-8 The three dimensional structure of phenylalanyl tRNA. *(Adapted from S.H. Kim et al., Science 185, 436 (1974)).*

the 5' to 3' direction), pairs with the phenylalanine codon UUC (again listed 5' to 3'). The reader should recall that polynucleotide pairing is always anti-parallel.

To synthesize proteins from amino acids, it is necessary to invest energy, since the hydrolysis of the peptide bond is energetically favorable, not its formation. It is in the formation of the complex between the tRNA and the amino acid, the amino acyl tRNA, that energy in the form of ATP is invested.

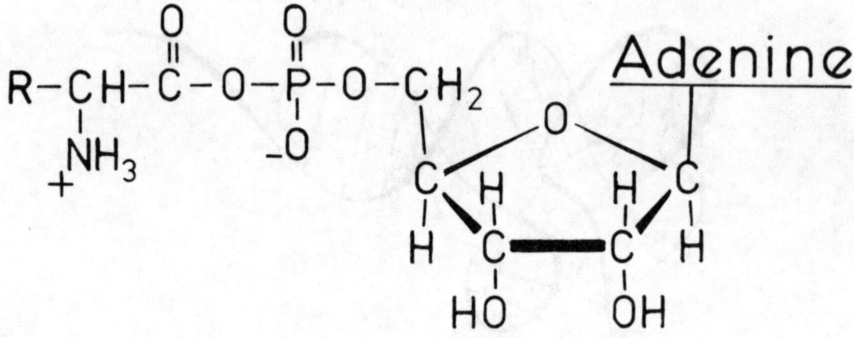

Fig. 6-9 An amino acyladenylate derivative of an amino acid. R stands for the side chain of one of the 20 amino acids for which codes exist.

First ATP reacts with an amino acid to give an amino acyladenylate derivative (Fig. 6-9) and pyrophosphate.

$$\text{Amino acid} + \text{ATP} \rightarrow \text{Amino acid-AMP} + \text{PP}_i; \quad \text{PP}_i \rightarrow 2\text{P}_i$$

The pyrophosphate is subsequently hydrolyzed with the result that this reaction is driven further to the right. The amino acid is then passed from the phosphate moiety of adenosine to the 3′ position on the ribose of the adenosine group at the 3′ end of the tRNA to form the amino acyl tRNA (Fig. 6-10) and AMP. Two high energy phosphodiester bonds have been hydrolyzed to form this tRNA-amino acid complex.

$$\text{Amino acid-AMP} + \text{tRNA} \rightarrow \text{Amino acid-tRNA} + \text{AMP}$$

Given an mRNA and a complete set of amino acyl tRNA's we are ready to assemble the protein. This process is catalyzed by a large macromolecular complex called the ribosome. These giant structures have molecular weights of

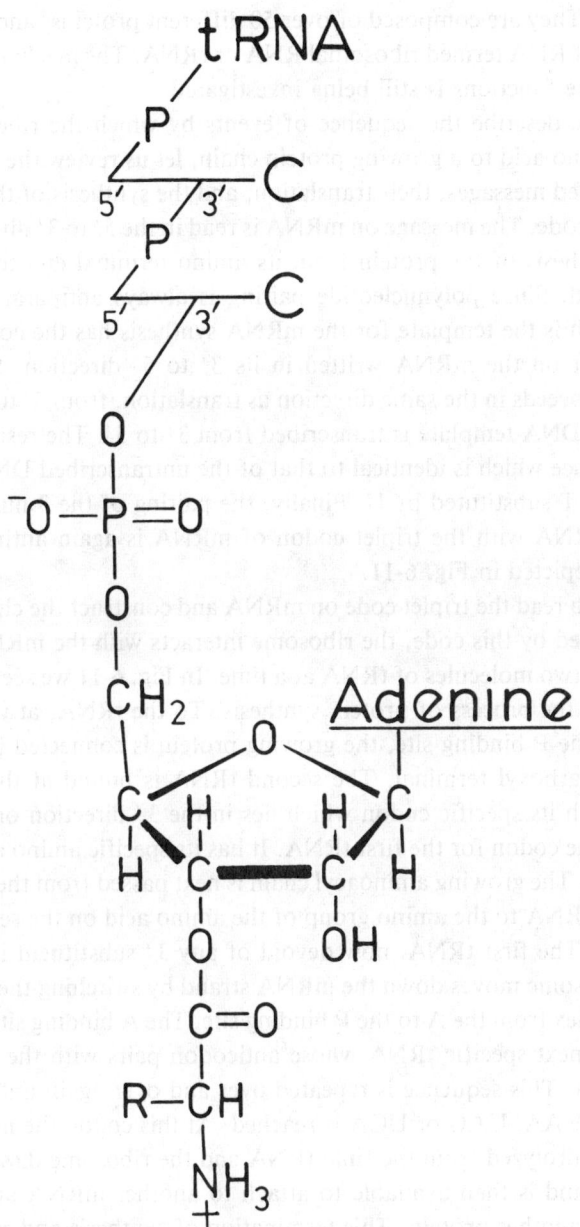

Fig. 6-10 An amino acyl tRNA.

some 3 million daltons in procaryotes and about 4 million in eucaryotic organisms. They are composed of over 50 different proteins, and a number of molecules of RNA termed ribosomal RNA or rRNA. The mechanism by which this structure functions is still being investigated.

Before we describe the sequence of events by which the ribosome adds a specific amino acid to a growing protein chain, let us review the directionality of these coded messages, their translation, and the synthesis of the protein for which they code. The message on mRNA is read in the 5' to 3' direction coding for the synthesis of the protein from its amino terminal end to its carboxyl terminal end. Since polynucleotide pairing is always antiparallel, the DNA strand which is the template for the mRNA synthesis has the complementary code to that on the mRNA written in its 3' to 5' direction. Since mRNA synthesis proceeds in the same direction as translation, from 5' to 3', the complementary DNA template is transcribed from 3' to 5'. The resulting mRNA has a sequence which is identical to that of the untranscribed DNA strand, of course with T substituted by U. Finally, the pairing of the 3 nucleotide anticodon of tRNA with the triplet codon of mRNA is again antiparallel. This pattern is depicted in Fig. 6-11.

In order to read the triplet code on mRNA and construct the chain of amino acids specified by this code, the ribosome interacts with the mRNA strand as well as with two molecules of tRNA at a time. In Fig. 6-11 we see a sketch of a ribosome in the process of protein synthesis. To the tRNA, at what we shall refer to as the P binding site, the growing protein is connected by the amino acid at its carboxyl terminal. The second tRNA is bound at the A site and interacts with its specific codon which lies in the 3' direction on the mRNA relative to the codon for the first tRNA. It has its specific amino acid attached at its 3' end. The growing aminoacyl chain is next passed from the 3' hydroxyl of the first tRNA to the amino group of the amino acid on the second aminoacyl tRNA. The first tRNA, now devoid of any 3' substituent is dissociated and the ribosome moves down the mRNA strand by switching the new protein tRNA complex from the A to the P binding site. The A binding site is now able to bind the next specific tRNA whose anticodon pairs with the next mRNA triplet codon. This sequence is repeated over and over again until a termination codon, UAA, UAG or UGA is reached. At this codon the now complete protein is hydrolyzed from the final tRNA and the ribosome dissociates from the mRNA and is then available to attach to another mRNA and begin the synthesis of another protein. This termination of synthesis and release of the finished protein requires the participation of other proteins, which again act as specific catalysts. It is important to note that the two sites on the ribosome

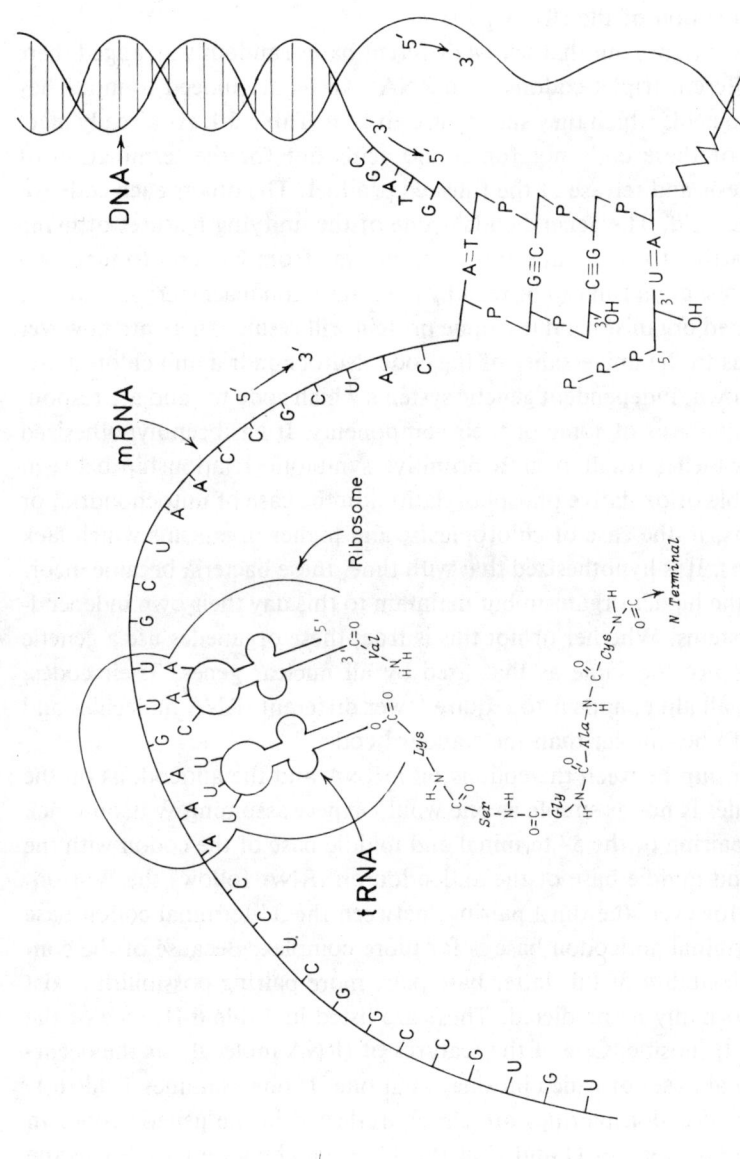

Fig. 6-11 Transcription and translation. The illustration depicts the transcription of one strand of a DNA molecule to form mRNA and the translation of that mRNA to synthesize a protein with a specific, defined sequence of amino acids. The directionality of all of these processes is indicated. Simultaneous transcription and translation as shown here could only occur in procaryotic cells since in nucleated cells these processes occur in different compartments.

which can be occupied by tRNA can accommodate any tRNA molecule. The specificity of binding is imparted entirely by the codon of the mRNA with which the anticodon of the tRNA pairs.

It is simple to figure out that with 4 different bases randomly arranged there can be 64 different triplet codons on mRNA. All 64 are indeed found. They and the amino acids which they specify are given in Table 6-I. As already mentioned three of these code not for amino acids but for the termination of protein synthesis and release of the finished product. The others each code for a single amino acid. The genetic code is one of the unifying features of living systems on earth. It is the same for all organisms from bacteria to man. An mRNA from one organism can be read by the translation machinery in another, totally unrelated organism and the same protein will result. There are however two exceptions to the universality of the code. Mitochondria and chloroplasts contain their own, independent genetic systems which code for and are responsible for the synthesis of some of their components. It has been hypothesized that these organelles result from a primitive symbiotic relationship between bacteria capable of oxidative phosphorylation, in the case of mitochondria, or photosynthesis, in the case of chloroplasts, and higher organisms which lack these functions. It is hypothesized that with time, these bacteria became incorporated into the higher organism but maintain to this day their own independent genetic systems. Whether or not this is true, these organelles use a genetic code which is not the same as that used by all nuclear genes. Their codes, which are not all alike, appear to require fewer different tRNA molecules and in that sense to be simpler than the standard code.

The relationship between the codons on mRNA and the anticodons on the tRNA molecules is not as simple as one would expect assuming Watson-Crick pairing. The pairing of the 5′ terminal and middle base of the codon with the 3′ terminal and middle base of the anticodon on tRNA follows the Watson-Crick rules. However, the third pairing, between the 3′ terminal codon base and the 5′ terminal anticodon base is far more complex. Because of the conformational flexibility of this latter base pair, more pairing possibilities exist than would normally be predicted. These are listed in Table 6-II. One of the bases listed is I, inosine. One of the features of tRNA molecules is the occurrence of unusual bases of which inosine is but one. If one examines Table 6-I, these unexpected codon pairings are clearly reflected in the genetic code. In every case the exchange of U and C at the 3′ terminal codon base leaves the message unaltered. In 14 out of 16 cases the exchange of A and G also has no result. In 8 out of 16 cases the message is completely independent of the nature of this 3′ terminal codon base. In one case, AUX, the message is the same

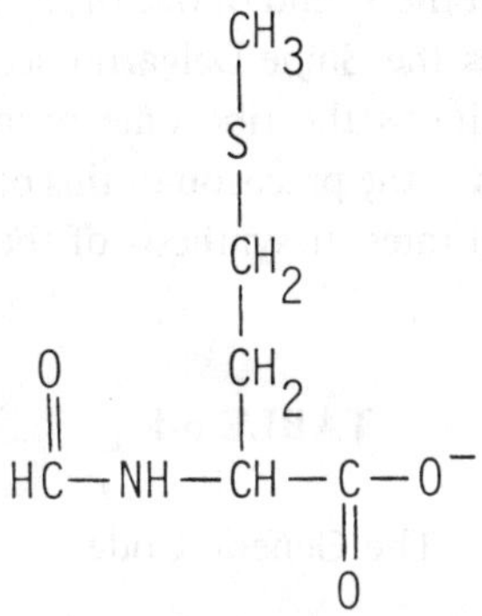

$$
\begin{array}{c}
CH_3 \\
| \\
S \\
| \\
CH_2 \\
| \\
\quad\quad\quad\quad CH_2 \\
O \quad\quad\quad | \\
\| \quad\quad\quad\quad | \\
HC-NH-CH-C-O^- \\
\| \\
O
\end{array}
$$

Fig. 6-12 Formylmethionine.

when X is U, C or A, but changes when X is G.

A critical step in protein synthesis has still not been discussed. This is the initiation process. Although the genetic code includes three stop codons which do not code for any amino acid, a similar set of start or initiation codons is not found. Instead, initiation occurs at a methionine codon, AUG on mRNA. Initiation requires a special methionyl tRNA. This tRNA binds directly to the P site on the ribosome. In procaryotic organisms the methionine bound to this initiator tRNA is chemically modified by linking a formyl moiety to its amino group (see Fig. 6-12). As a result the amino acid sequences of newly synthesized bacterial proteins always begin with formylmethionine. In eucaryotic organisms the initiator methionyl tRNA is not formylated so newly synthesized proteins in higher organisms begin with methionine. Proteins frequently undergo post synthetic modifications which often include removal of their N-terminal methionyl or formyl methionyl residues. Therefore, it is not true that all proteins in their finished form begin with a derivative of methionine.

The next question we must consider is how the translation apparatus knows at which AUG codon to begin. Again the answer differs between procaryotic and eucaryotic organisms. In the latter case, the mRNA normally contains the information for the synthesis of only one protein, and initiation occurs at the AUG codon closest to its 5 ' end. In procaryotes, where mRNA molecules frequently code for multiple proteins, the situation is more complex. In such organisms one finds a purine rich sequence centered about 10 nucleotides 5 ' from the initiator codon. This is called the Shine-Dalgarno Sequence, after its discoverers. There is a pyrimidine rich sequence at the 3 ' end of one of the ribosomal RNA molecules which apparently recognizes the Shine-Delgarno sequence by complementary base pairing, and thus positions the ribosome properly for the initiation of protein synthesis. Variations in the precision of this base pairing may explain, at least in part, the differential rates of synthesis of the various proteins of these organisms.

TABLE 6-I

The Genetic Code

3 ' Terminal Base	Middle or Second Base				5 ' Terminal Base
	U	C	A	G	
U	Phe	Ser	Tyr	Cys	U
	Phe	Ser	Tyr	Cys	C
	Leu	Ser	Stop	Stop	A
	Leu	Ser	Stop	Trp	G
C	Leu	Pro	His	Arg	U
	Leu	Pro	His	Arg	C
	Leu	Pro	Gln	Arg	A
	Leu	Pro	Gln	Arg	G
A	Ile	Thr	Asn	Ser	U
	Ile	Thr	Asn	Ser	C
	Ile	Thr	Lys	Arg	A
	Met	Thr	Lys	Arg	G
G	Val	Ala	Asp	Gly	U
	Val	Ala	Asp	Gly	C
	Val	Ala	Glu	Gly	A
	Val	Ala	Glu	Gly	G

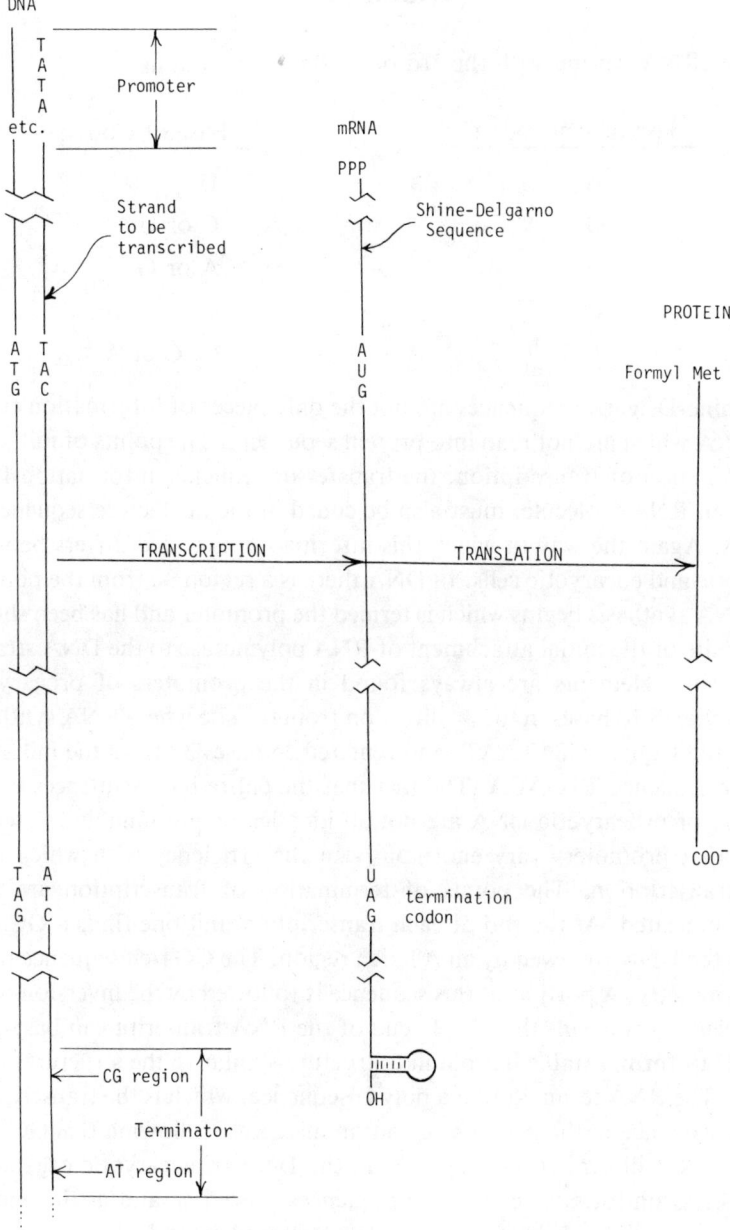

Fig. 6-13 The locations and patterns of nucleotides which code for the beginning and end of transcription and translation in procaryotic cells.

TABLE 6-II

Rules for tRNA pairing with the 3rd or 3 ′ Bases of Codons

5 ′ Base of Anticodon	3 ′ Base of Codon
A	U
G	C or U
U	A or G
C	G
I	U, C or A

The Shine-Delgarno sequences are not the only pieces of information coded in the DNA which are not read into protein sequences. The points of initiation and termination of transcription, the transfer of sequence information from DNA to an RNA molecule, must also be coded in the nucleotide sequence of the DNA. Again the way in which this information is coded differs between procaryotic and eucaryotic cells. In DNA there is a region 3 ′ from the point at which RNA synthesis begins which is termed the promoter and has been shown to be the site of the initial attachment of RNA polymerase to the DNA strand. Two sequence elements are always found in the promoters of procaryotic DNA. Centered 10 bases in the 3 ′ direction from the site where RNA synthesis begins is the sequence TATAAT, and centered 35 bases 3 ′ from the initiation site is the sequence TTGACA. The fact that the entire base sequences of the promoters of procaryotic DNA are not all identical is presumably related to the fact that promoters vary enormously in the efficiency with which they initiate transcription. The points of termination of transcription are also precisely indicated. At the end of each transcription unit one finds a GC rich region in the DNA followed by an AT rich region. The CG rich sequence has a special symmetry. A portion of this sequence is followed by the inversion of its complement. This means that the 3 ′ end of the RNA transcript can base-pair with itself to form a stable hairpin like structure similar to the structures seen in tRNA. The RNA terminates in a poly U sequence, which is the transcript of a poly A sequence in the AT rich region, immediately following this GC rich hairpin. This is illustrated in Fig. 6-13. In the DNA of eucaryotic organisms one finds less uniformity in the base sequences preceding and at the end of transcription units, and these codes are less well understood.

The initiation of the transcription of the information from DNA onto the RNA molecule is one of the key steps in the control of protein synthesis. As has already been mentioned, the rate of initiation appears to vary with the sequence of bases in the promoter sequence. Furthermore, this rate can be modulated by proteins which inhibit (repressors) or augment (activators) the binding of the RNA polymerase to the promoter or its movement down the DNA template. An example of such control is seen in the lac operon, the transcription unit in the bacterium, *E. coli*, which contains the genes which code for the enzymes required for lactose metabolism. Lactose is a complex of two hexose sugars, one glucose and one galactose molecule. In the absence of lactose there is little reason to utilize energy in synthesizing these enzymes, and indeed they are produced only in very small amounts. However, if the organisms are placed in a medium containing lactose as the principle energy source these enzymes are produced in large amounts. We now know that another gene termed the i gene which lies near the lac operon but which is in a separate transcription unit, codes for a protein which binds specifically at a site on the DNA, called the operator, which is adjacent to or overlapping with the lac promoter sequence. The binding of this repressor protein to the operator site inhibits transcription. The addition of lactose to the medium results in the formation of allolactose by the small amounts of the necessary enzymes that are present even in the absence of lactose. Allolactose binds tightly to the repressor protein and inhibits its interaction with the operator site thus permitting transcription to proceed.

The control of gene expression in response to environmental changes is obviously of great value to an organism. However, in higher organisms gene expression is modulated by a variety of factors some of which appear to be defined functions of time and are the secret to how a single cell can multiply and its progeny differentiate in such a controlled fashion as to produce a functioning human being. The modulation of gene expression in differentiation is not understood, but there is now considerable evidence that it involves chemical modification of the genome. One form of modification is the reversible methylation of nucleotide bases. However, there is also evidence, particularly in the differentiation of antibody producing cells, of gross rearrangements of the genome and the apparent destruction of genetic elements. If the latter indeed occurs it would seem to violate the central dogma which states that information transfer from DNA to proteins is unidirectional. However, the reason for this stipulation is that the DNA must remain unmodified so that it can code for future generations of organisms. In single-cell organisms this is a fundamental requirement. However, in complex organisms such as vertebrates there is a precise segregation of germ line cells, from which future generations arise, and somatic cells all of which are members of cell lines which terminate with the death of the organism. Since the latter cells will not furnish the genetic information for the production of new organisms, their genetic material need not be immutable.

We have discussed the mechanism by which the nucleotide sequence in the DNA molecule codes for the amino acid sequences of all of the proteins in the organism. The next question with which we must concern ourselves, is how is the information contained in the sequence of amino acids in a protein utilized? An amino acid sequence is not a linear code for some other structure. In fact, there is no mechanism by which it can be used to code for an RNA or DNA sequence appropriate for its own duplication. Instead, the sequence of the amino acids in a protein codes for the final three dimensional structure and functional properties, such as catalytic activity, of the protein. The code of a nucleotide sequence is read by a large macromolecular complex, the ribosome in conjunction with the set of tRNA molecules. The information in an amino acid sequence is read by the physical forces that exist between chemical residues or groups in the aqueous milieu of living systems as we know them. That is, the folding of a protein into its native or functional structure is a spontaneous process. The final product of which does not depend on specific catalysis by any other substance.

In a polymer of the twenty different amino acids, for which there are codes in mRNA, there are numerous possibilities for energetically favorable interactions both among the different groups of the protein and between these groups and molecules of solvent water. The peptide backbone of the protein is a series of amide linkages which are very polar structures. They can take part in hydrogen bonding interactions with water molecules or with other groups on the protein. Some repetitive structural forms of polypeptides involve H-bond interactions between these peptide elements. An example of one such structure, the α-helix, is shown in Fig. 6-14. The 20 amino acids present a wide variety of side chains with numerous possibilities for interactions (see Fig. 6-5). At normal physiological pH some side chains are negatively charged (glutamic acid, aspartic acid) while others carry positive charges (arginine, lysine). Many side chains are polar moieties (serine, glutamine, etc.) while others are non-polar or hydrophobic (leucine, methionine, phenylalanine, etc.). The thiol groups of the cysteine residues can be oxidized to disulfide bonds to give a covalent cross link between residues of a peptide chain. In addition, favorable interactions occur among the water molecules of the solvent. The energetics of the folding of any polypeptide can be determined only by summing the effects that the particular folding process has on all of these interacting systems.

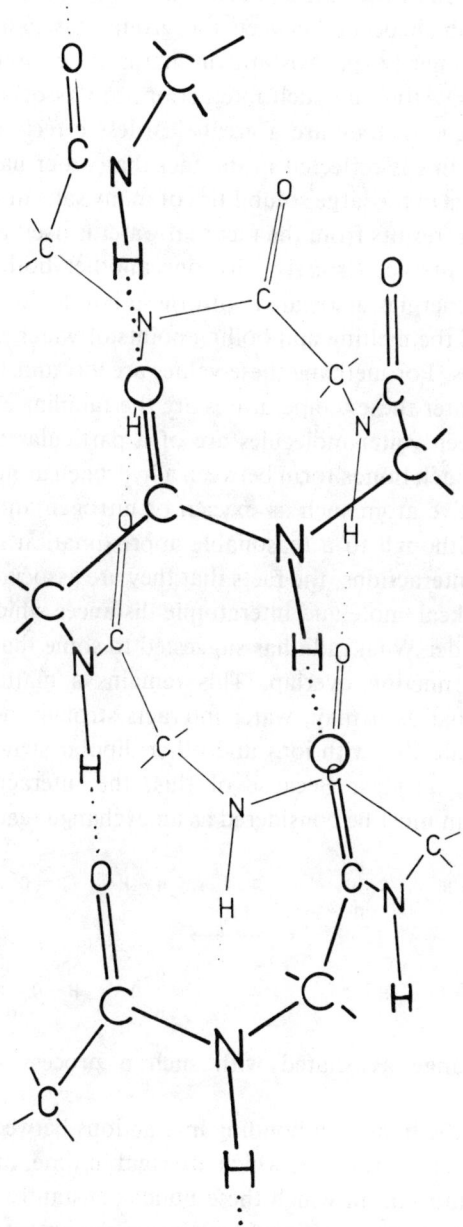

Fig. 6-14 The three dimensional structure of the peptide backbone of an α-helix.

With the exception of the covalent disulfide bond between cysteine residues, the interactions which occur between the groups discussed in the previous paragraph are primarily electrostatic in nature, i.e. interactions between charges or dipoles. Although such interactions are associated with considerable energy in vacuum, they are a great deal less effective in water. At the macroscopic level this is reflected in the fact that water has a large dielectric constant and results in the large solubilities of many salts in this solvent. At the molecular level this results from the fact that water is itself a dipolar molecule. Water molecules interact strongly with one another both in ice and in the liquid state. The energies associated with these interactions are suggested by the comparison of the melting and boiling points of water and methane which are of similar mass. For methane these values are 90° and 112° Kelvin respectively while for water these temperatures are the familiar 273° and 373°. The interactions between water molecules are of a particular type referred to as hydrogen bonds. Such bonds form between a hydrogen atom covalently bound to an electronegative atom such as oxygen or nitrogen and a second electronegative atom. Although to a reasonable approximation such bonds can be treated as dipole interactions, the facts that they are associated with enthalpies of the order of 5 kcal/mole and interatomic distances which are shorter than predicted by Van der Waal radii has suggested to some that they may involve significant wave function overlap. This remains a matter of considerable controversy. Be that as it may, water interacts strongly not only with other water molecules but also with ions and other dipolar structures, particularly hydrogen bonding groups. Because of this, the interaction between such groups on a protein must be considered as an exchange reaction, for example,

The enthalpy change associated with such a process need not be very significant.

In liquid water the hydrogen bonding interactions between water molecules obviously do not form structures which are rigid in time, and must instead be in a dynamic equilibrium in which these bonds constantly form and rupture. However, at any instant a significant fraction of the potential hydrogen bonding groups must be involved in such interactions. This dynamic equilibrium allows a water molecule considerable rotational freedom since potential hydrogen bonding groups on other water molecules are to be found in all directions. A very different situation is found at the surface of a volume of water. Water molecules at the surface are faced with a large solid angle in which there are no hydrogen bonding groups with which to interact. There are two possible responses to this situation. The water molecules can maintain their orientational freedom, sacrificing a certain number of hydrogen bonding interactions in the process. In this case the surface will be an enthalpically unfavorable location. Alternatively hydrogen bonding interactions can be maintained by limiting the rotational freedom of the surface molecules, resulting in an entropically unfavorable state. In either case the free energy of the water molecules at the surface will be higher than that of molecules in the bulk volume. This is the reason for the large surface tension exhibited by water. The free energy of a volume of water is reduced by reducing the area of such surfaces. At normal physiological temperatures this free energy change is primarily entropic. The surface we have described is the interface of water with any non-polar volume. It can be an air-water interface or, of more interest to us in the present context, the interaction of water with a non-polar compound or chemical group such as a hydrocarbon. This is the basis for the "hydrophobic interaction", the tendency of non-polar groups or substances to coalesce when immersed in water. In Table 6-III are presented the thermodynamic changes which result when a series of non-polar molecules are transferred from a non-polar medium to water. One observes that such transfer

TABLE 6-III

Thermodynamics properties of the transfer of non-polar solutes from non-polar solvents to water*

REACTION		TEMPERATURE (degrees Kelvin)	ΔS (e.u.)	ΔH (cal/mole)	ΔG (cal/mole)
CH_4 in benzene	CH_4 in water	298	-18	-2800	$+2600$
CH_4 in ether	CH_4 in water	298	-19	-2400	$+3300$
CH_4 in CCl_4	CH_4 in water	298	-18	-2500	$+2900$
C_2H_6 in benzene	C_2H_6 in water	298	-20	-2200	$+3800$
C_2H_6 in CCl_4	C_2H_6 in water	298	-18	-1700	$+3700$
C_2H_4 in benzene	C_2H_4 in water		-15	-1600	$+2900$

* Data from Kauzmann, W. (1959) Adv. Prot. Chem. 16:1–63

processes are invariably associated with negative values of ΔS. Surprisingly, for many of these processes ΔH is strongly negative. It is doubtful that this arises from interactions between the non-polar solute and water, and it has been suggested that water molecules can form clathrate shells around such solutes resulting in more complete hydrogen bonding than occurs in the bulk phase.

Such hydrophobic interactions appear to control the general pattern of folding of proteins in water. This conclusion rests on the observation that in their native structures proteins invariably are folded in such a way as to place their hydrophobic groups into their interior spaces away from solvent water, while their surfaces are occupied by polar or charged moieties. The occurrence of unique, well defined structures for biological proteins cannot be explained by hydrophobic interactions alone because these effects lack specificity. There are no orientational requirements placed on the interacting, non-polar residues, since their interactions depend not on mutual attraction but on essentially a repulsion by water. However, in removing non-polar groups from exposure to solvent many polar moieties of the protein are denied contact with water. This would be energetically unfavorable were not other energetically favorable interactions established in the interior of the protein. It is these interactions which have strict geometric requirements and confer specificity on the folding process. This is a rather incomplete treatment of this problem. However, it should be sufficient to make the point that the folding of a protein results from a balance between competing reactions or processes. It involves the rupture of favorable protein-water hydrogen bonding interactions with the formation of protein-protein hydrogen bonds, and a favorable change in the entropy of the solvent which is opposed by the unfavorable entropy effect of conferring on the protein a single conformation.

That these forces, acting directly on the polyamino acid chain, without intervening catalytic activities, determine the native structure and biological activities of proteins has been amply demonstrated. The classic studies were carried out with the enzyme ribonuclease which catalyzes the hydrolysis of the phosphodiester linkages between the ribose sugar moieties of RNA. This enzyme is a single polypeptide chain of 126 amino acids. It contains 4 disulfide bonds between cysteine residues. If these disulfide bonds are reduced to form a set of unlinked cysteine residues and the protein placed in a solvent which disrupts hydrophobic interactions and hydrogen bonds, a solvent such as 8M urea in water, the protein is converted to a structureless randomly oriented extended polypeptide chain. If the urea is slowly removed from the system and the disulfide bonds permitted to slowly reform by oxidation, this protein will refold into its native structure with full enzymatic activity. Similar experiments have been carried out for many proteins with similar results.

For some proteins such refolding experiments do not yield the native biologically active structure. One example of such a protein is the hormone insulin. This protein is composed of two polypeptide chains which are linked by disulfide bonds. Once dissociated, these two chains seem to have a low probability of reassociating in the proper orientation to reform the native structure. However, it is now known that insulin is originally produced as a single chained, biologically inactive protein, proinsulin. Proinsulin can be renatured from its unfolded state. Proinsulin is activated by proteolytic cleavage which converts it to the double chained insulin molecule. Such activation by selective hydrolysis of certain peptide bonds in a protein represents still another mechanism for the control of the enzymatic or other biological activities of proteins.

Since the code of the amino acid sequence of a protein is read by the physical forces which exist between chemical groups in solution, it should be possible for man to learn to read this code and to predict the structure of a protein from its amino acid sequence. To date attempts to do this have not been particularly successful. This reflects a number of problems in carrying out such computations. These include the enormous number of variables which must be fitted, and the limitations in the precision of our knowledge of the energetics of the interactions which occur in solutions of electrolytes. Additionally, in most attempts to do this it has been assumed that the final structure of the protein represents the equilibrium structure, i.e. the structure of lowest free energy. Considerable evidence has now accumulated to suggest that this may not be the case. If one makes even modest estimates of the number of different conformational states in which a protein of say 100 amino acids can exist and then assumes that the rate of sampling of these states is limited by the vibrational frequencies of chemical bonds, one finds that the time required for the folding of a protein into its native state is too short to permit the sampling of all possible states. If this is true it precludes that the system is at equilibrium. It has been suggested that protein folding begins with a series of nucleation events in which structures are formed which involve interactions between residues which are in close proximity in the amino acid sequence. Interactions among these local structures then lead to the final conformational state. If such folding pathways do exist, then the structure of a protein probably represents only a local energy minimum and may not be the state of lowest free energy. If this is so then the structure can be computed only by including kinetic considerations in one's calculations.

The fact that the amino acid sequence of a protein defines a single three dimensional structure, or perhaps more accurately a limited population of closely related structures, is of great importance to living systems. Were proteins to fold up into an array of different structures, only a few of which were biologically active, one would have a highly inefficient system. It is very doubtful that having a single structure is a property of all or even most polypeptides, but for biologically useful proteins it would seem to be necessity. This, along with the biological activity of the favored structure must be a major factor in the evolutionary selection of protein sequences.

The proteins thus coded for by the DNA of an organism are responsible directly or indirectly, for all of the processes required by the organism. They synthesize, through specific catalysis, all of the compounds required by living systems including nucleotide triphosphates and from them DNA and RNA, and amino acids and from them other proteins. Frequently their catalytic activities require the participation of non-protein moieties from something as simple as a metal ion to complex organic compounds such as the hemes found in the cytochromes. However, in all cases the proteins confer on these compounds unique properties which permit them to carry out specific chemical roles. All of the organic compounds of living systems are the result of the catalytic activity of proteins. True, organisms such as man ingest large quantities of complex organic materials and use them both as energy sources and as constituents of essential structures. Some of these ingested compounds are essential components of our diet in that we are unable to synthesize them ourselves. These include the vitamins. However, all of these compounds are ultimately synthesized by some living organism using the specific catalytic activities of proteins. Proteins also serve a structural role. The tendons that attach our muscles to our bones are composed of the structural protein, collagen. In muscle proteins serve as both the structural elements and the catalytic components which convert chemical energy into mechanical work, motion.

There is one other class of biological polymers which deserves some mention. This is composed of the polysaccharides, the polymers of sugar residues. We have already discussed glycogen in Chapter 4. This polymer of glucose differs from cellulose, the structural polysaccharide of plants, only in the stereochemistry of the linkage between the glucose moieties. Such uniform, repeating polymers serve a structural or storage function and contain essentially no information. However, complex polymers of different sugar residues do occur in which the sequence of the sugars and the linkages between them are precisely defined. Unlike proteins and nucleic acids these complex polysaccharides are not the result of a linear translation of a code on another molecule. Instead they result from the activities of a set of enzymes, glycosyl transferases, the specificity of each of which is such that it can attach a particular sugar to a particular sugar by a particular stereochemical linkage. The final structure obtained depends on which enzyme activities are present. Such complex polysaccharides are commonly found attached to proteins on the surfaces of cells, and are used among other things for recognition. As one might expect, the molecules which recognize these structures by specific binding are other proteins.

The structures which we have discussed in this chapter are all of importance in information storage, retrieval and processing. Nucleic acids store a linear code which can be read into the structures of proteins. Many of these proteins possess binding sites which can be of exquisite specificity. The structure of a protein results from the balance of forces or energies of the interactions among the chemical residues of the protein and between these residues and the solvent water. The binding of a substance at a specific binding site can alter this balance and thus change the structure of the protein. Such a change in structure can be detected by other specific interactions, and result in an altered enzyme activity. This yields a pattern of recognition and response. This is the basic mechanism of hormone action as exemplified by the activation of adenyl cyclase on the inside of a cell by the binding of epinephrine to its exterior (see Chapter 5). Such recognition and response by protein systems also play a role in vision and neurotransmission. Information transfer and processing, i.e. communication, is one of the most fundamental features of higher life forms. Much of this, such as the functioning of the brain, remains poorly understood. Nevertheless, it is clear from the material reviewed in this chapter that a good beginning has been made in understanding some of these phenomena at the molecular level.

FURTHER READINGS

General

Watson, J.D. (1975) *Molecular Biology of the Gene*, Benjamin.

Freifelder, D. (1978) *The DNA molecule: structure and properties*, Freeman.

Cantor, C.R. and Schimmel, P.R. (1980) *Biophysical chemistry (Part I) The conformation of biological macromolecules*, Freeman.

Cantor, C.R. and Schimmel, P.R. (1980) *Biophysical chemistry (Part III) The behavior of biological macromolecules*, Freeman.

Specific

Zimmerman, S.B. "The three-dimensional structure of DNA", (1982) *Annu. Rev. Biochem.* **51**:395.

Kornberg, A. (1980) *DNA replication*, Freeman.

Rich, A. and Kim. S.H. "The three dimensional structure of transfer RNA", (1978) *Sci. Amer.* **238**(1):52.

Lake, J.A. "The ribosome", (1981) *Sci. Am.* **245**:84.

Crick, F.H.C. "The origin of the Genetic code", (1968) *J. Mol. Biol.* **38**:367.

T. Hunt "Peptide chain termination", (1980) *Trends Biochem. Sci.* **5**:234.

CHAPTER 7

ORIGIN OF LIFE: FACTS, HYPOTHESIS AND MODELS

Cosmologists are often wrong, but never have doubts.

Lev Landau

The origin and evolution of living systems on earth became a matter of science only in the middle of last century. Studies in this area have frequently involved the concepts of physics. Even fundamental laws of physics have been used to criticize or to support various theories. When Charles Darwin proposed his hypothesis of biological evolution, he was authoritatively contradicted by the leading physicist Lord Kelvin, who found Darwin's proposal to be manifestly untenable on grounds of the thermal physics of the sun and of the earth: the sun and the earth would have not been able respectively to shine and to keep warm enough for the number of years in the past required by Darwin's evolutionary theory. Reportedly, some of Darwin's followers were very worried, by this apparent clash with the "fundamentals" of physics; thermodynamics was and is a theory of great appeal in its simplicity and elegance and, of course, fully supported by all known experimental situations. Darwin, on the other hand, thought it possible that physicists would change their minds and so told his collaborators not to worry too much. And physicists did actually change their minds, not about thermodynamics of course, but about the burning processes in the sun and the cooling of the earth. The problem was that Kelvin, from the evidence available at that time, had to believe the sun was burning on chemical energy rather than on nuclear reactions as we now know. Also, having no notion of radioactivity, he did not take into account the slowing down in cooling of earth due to internal heating resulting from the radioactivity of rocks.

The Table 7-I shows that there is no longer any conflict between characteristic times of cosmological and biological evolution.

TABLE 7-I

Estimated ages (years)

universe	$10 \div 20 \times 10^9$
sun	5×10^9
earth	4.5×10^9
oldest earth rocks	$> 3.5 \times 10^9$
microfossils	3.2×10^9
blue green algae	2.3×10^9
eucaryotes	1.3×10^9
hominid footprints	3.6×10^6

Physics at its birth with Copernicus and Galileo denied us of the privilege of being at the center of the universe. Now from knowledge of the physical processes occurring in stars it seems that another privilege is denied to us: that of being somehow unique. The number of stars much like our sun is large. Although astronomers are only beginning to obtain direct observational evidence of planetary systems in nearby stars, it still seems to be a good guess that most stars have planets. A conservative estimate based on counts of galaxies and stars and on considerations of stellar evolution suggests that in our universe some 10^{18} locations could be more or less suitable to the development of living structures, in the sense that conditions may be similar to those on our earth. By this estimate there should be about 15 planets "suitable for life" within 20 light years of us. The time interval in which living structures which develop technologies can be active may be very short in respect to the planetary evolutionary times. Hoyle estimates 3×10^5 years to be the time interval in which a "technology" is active before destroying itself by unwise use of the gadgets it has created. This last remark may be particularly appropriate, as in these days (spring 1984), the so called superpowers are unable even to start a deal for the reduction of nuclear armaments. The tragicomical side of the story is that, in case they would start such a deal, the goal would be to reduce nuclear weapons from ten times to six times as much needed to destroy all life on Earth! If one accepts these estimates then at the present epoch some 6×10^7 technologies should be active in the universe at the average distance of 200 light years. One could try to communicate. To do so one should send and attempt to receive messages above the background noise, without knowing where the party is, on which frequency he is broadcasting and what code is being using. This is possibly an overwhelming problem because of the high cost of any reasonable effort and it is a matter of debate whether it can be usefully attacked. Therefore, even if other civilizations exist in outer space, we may have very little chance of ever communicating with them.

Given that biological evolution is not in conflict with the known physics of the evolution of the universe and that it is actually quite convincingly supported by systematic studies of the fossil records, as we shall see below, the next problem is how it all started.

Here one confronts a serious problem for which there is no easy solution. We have evidence for the occurrence of only one such event, the origin of life here on earth. We have no way of knowing whether the time required for the appearance of life following the formation of the earth represents a roughly average time for such an occurrence in that environment or whether it represents a rare phenomenon, unique in the universe. The problem is compounded by the fact that we are not neutral observers of this happening since the creation of life is a necessary condition for our existence. Therefore, even if it had occurred only once in the history of the universe, it is that single event of which we would be aware.

It is therefore difficult to argue against the view that a highly improbable fluctuation gave rise initially to a living system by creating a unique set of conditions at a favorable moment during the evolution of the earth. Thereafter the known laws of physics and chemistry have been regulating the biochemical and biological evolution we participate in. If this were so then we would be continuously threatened by disaster, which eventually will cancel life from the universe, because sudden changes in environmental conditions, which could destroy all life on earth, should be much less improbable, than the favorable fluctuation which created it.

However, this position leads nowhere. Furthermore, one can imagine numerous steps in the formation of living systems, not just an initial one, which might be highly improbable. Therefore, it seems more useful to make the opposite hypothesis, that the time required for life to form on the earth was a roughly average time for such a process. Given this hypotheses, one can then explore the extent to which individual steps in this process can be reproduced in the laboratory or explained convincingly.

Another view, which was common in the middle of this century, held that the present laws of physics would be inadequate to explain life; new basic laws of nature would be needed. We cannot at present prove this position to be false. However, as more and more properties of living systems are explained by the known laws of physics, the need for new laws appears less and less likely.

From the time scale shown in Table 7-I, it seems that life appeared on our planet very soon after its accretion from the nebula, which gave rise to the solar system. If this represents an average time for this process then it appears that the generation of living structures is a much more likely phenomenon, than their present complexity would suggest. It seems possible that the progressive synthesis of more complex molecular structures and their orderly aggregation in precursors of living systems could be a compulsory rather than exceptional pathway in the generalized cooldown of our expanding universe, just as, from primordial hydrogen and helium, the present variety of elements has been created through the evolution of galaxies and stars.

The search for the origin and evolution of living systems from simple chemicals is very analogous to cosmology in its overall methods. The approach is interdisciplinary, from astrophysics to geochemistry to paleobiology, to information theory, to non-linear mathematical physics and chemistry. It is obvious that the biggest problem is prebiotic evolution, since Darwin's principles of natural selection will account for the subsequent evolution of any initial population of living structures. At the same time it is to be hoped that the description of such prebiotic evolution will offer a physical basis for the rules of natural selection, so that this property of biological systems can be reduced to a fundamental law of nature.

The approach is multifaceted and includes i) reconstruction of the history of the earth from its formation, ii) determination, through a painstaking search of records, of the occurrence, location, and the history of organic, biochemical and biological structures on the earth, iii) attempts to reproduce in the laboratory at least some of the paths of chemical evolution relevant to living systems, iv) attempts to advance and test in the laboratory reasonable models which predict "spontaneous" evolution from a mixture of chemicals to a culture of aggregates which duplicate, mutate, compete.

Direct methods of dating rocks rely on measurements of their radioactivity at present and the relative abundancies of one or more pairs of parent A and product B radioactive nuclides. The "birth" of the rock is assumed to have taken place at time T in the past when the radioactive nuclide A went into a solid phase so that it could no longer be segregated by diffusion and/or chemical reactions from its product B. To date ancient rocks one must obviously chose radioactive nuclides with decay time $1/\gamma$ larger than T. One must also assume that before the time T the product B was promptly separated in the melted phase from the parent A, so that at the initial time the rock contained no B nuclides and n_A A nuclides, which started decaying according to $n_A(t) = n_A e^{-\gamma t}$. At present, after the time T elapsed, we have then that the following relation holds between the relative abundancies of A and B

$$\frac{n_B}{n_A} = e^{\gamma T} - 1 .$$

Therefore a measurement of n_B/n_A and the knowledge of the decay constant γ permits one to determine T. The actual situation in case one is trying to date very old rocks, is a bit more complicated because in some relevant cases the product can be a gas present in the atmosphere or the parent may have more than one decay mode or other diffusive processes may weaken the fundamental hypothesis of complete segregation of the radioactive parent nuclides. So usually one attempts to obtain concordance between estimates of T from several pairs of nuclides. The Table 7-II gives a summary of the nuclides most

TABLE 7-II

Nuclides for radioactive dating

$A \rightarrow B$	γ 10^{-10} (years)$^{-1}$
$Rb^{87} \rightarrow Sr^{87}$	0.14
$Re^{187} \rightarrow Os^{157}$	0.16
$Th^{232} \rightarrow Pb^{208}$	0.50
$K^{40} \rightarrow Ar^{40}$	0.58
$U^{238} \rightarrow Pb^{206}$	1.5
$U^{235} \rightarrow Pb^{207}$	9.7

frequently used for these measurements and of their relevant properties. The most ancient rocks found on the Earth's crust have been directly dated to about 3.5×10^9 years ago. There are reasons to believe that rocks found in the Brazilian island of Sao Paulo have remained unaltered for the last 4.5×10^9 years, although, because of their low content of radioactive nuclides, it has been impossible to obtain a direct dating. Making the additional hypothesis that the solid objects in the solar systems are all segregated and condensed in a time which was short in respect to the Earth's age, no matter what the detailed model chosen for this process, then one would guess that the best estimate for the Earth's age comes from the study of meteorites, as they have always been well isolated in space. Direct dating methods applied to meteorites give $T = 4.5 \times 10^9$ years. Lunar rocks agree with this figure, which is presently taken to be the age of Earth.

In a model, according to which the Earth was accreted from grains, dust and gas it seems that any gas would have been gradually dispersed and that the primitive solar nebula could not have contributed to the early Earth atmosphere. The very primitive atmosphere of Earth was then formed by the outgasing and volatilization of impacting meteorites and was constituted of H_2O, CO_2, N_2 and noble gases. The whole process of collapse of the primitive solar nebula, accretion of the Earth, dissipation of the remnants of the nebula and formation of such a primitive atmosphere could have occurred in a time of as little as 10^8 years, during which the newly formed Sun would have moved into the main sequence of stellar evolution. The photochemical processes occurring in such a situation would have produced formaldehyde, a key precursor of organic molecules.

According to a different point of view, which was initiated by Oparin and has greatly influenced many successful attempts to test primordial chemical synthesis in the laboratory, the atmosphere of the early Earth was highly reducing and was composed of hydrogen, methane, ammonia, and water, as is found for the largest planets, Jupiter and Saturn, which have retained their early atmospheres. Again the chemical reactions promoted in a properly hot and sparky environment would yield significant quantities of organic molecules.

The search for fossils of the most ancient living creatures has given interesting results. Microfossils of spheroidal shape, 2.5 μm in diameter have been found in sediments of the Archean Swaziland System in South Africa, which are believed to be over 3×10^9 years old. The walls of these microstructures show the presence of organic chemicals and those found in binary aggregation, show the morphology of present day procariotes in the process of division,

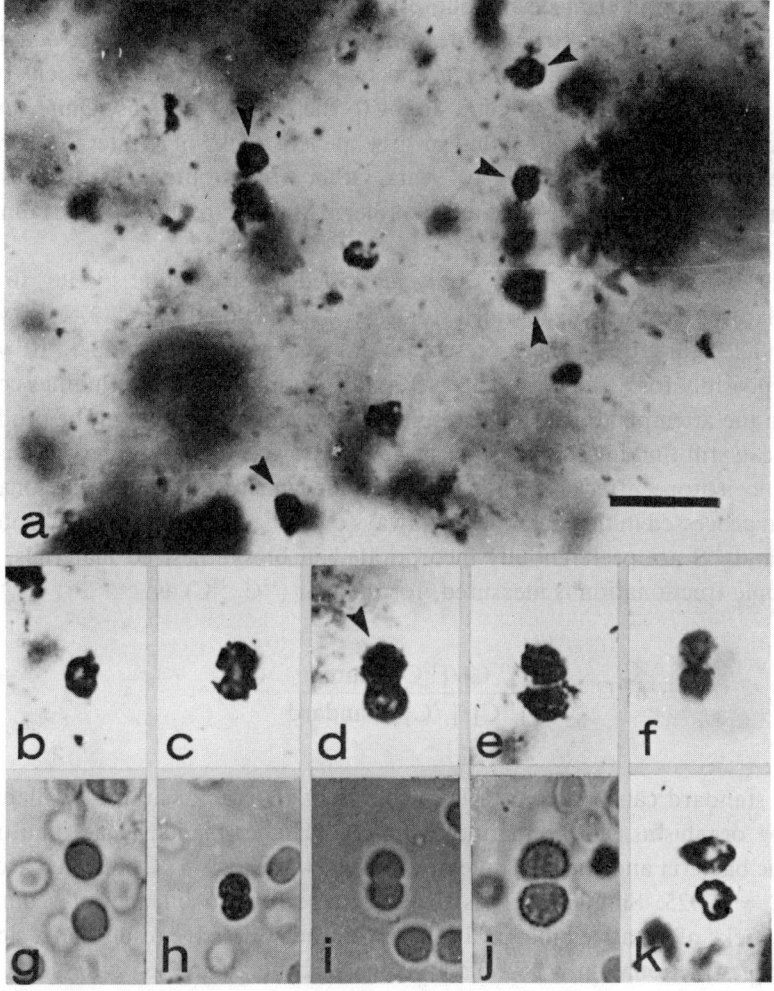

Fig. 7-1 (a) Distribution of microfossils in carbonaceous laminae of the Swartkoppie Formation, South Africa; arrows designate discernible individuals.

(b to e) Stages in cell division preserved in the Swartkoppie population; the arrow in (c) points to dark organic contents within the upper half of the dividing cell.

(f and k) Dyads in the Swartkoppie population.

(g to j) Aphanocapsa sp., illustrating binary cell division in modern prokaryotes. All photographs, × 1120.

*(Reproduced with permission from: A. H. Knoll and E. S. Barghoorn, Science **198**, 396 (1977)). Copyright © (1977) American Association for the Advancement of Science.*

Fig. 7-1. Some green algae and photosynthetic bacteria produce laminar strati-fied sediments called stromatolites, as a result of their metabolic activities. Various morphologies are known and have been recognized in recent rocks and shown to be related beyond doubt to these living systems. Quite similar structures have now been found in very ancient rocks in Western Australia, believed to be as old as 3.5×10^9 years. Other stromatolite-like structures had been found at sites in Canada and Southern Africa in rocks with ages ranging from 2.5 to 3.0×10^9 years.

The oldest microfossils positively identified as procariotic species, in fact similar to blue green algae which still exist, date back to about 2.3×10^9 years ago. Therefore life has been present on this planet for at least 2.3×10^9 years and possibly for as long as 3.5×10^9 years. Support for these findings comes from the attempts to detect the results of biochemical activity in the products one can still find in sediments. As has been known for a long time, because of kinetic isotope effects, carbon and sulphur undergo isotopic fractionation when processed in the metabolic pathways of living systems. The light isotopes ^{12}C and ^{32}S are preferentially incorporated in biosynthesized materials. The isotopic fractionation is measured, for the pair (^{13}C, ^{12}C) as

$$\delta^{13}C = [\frac{([^{13}C]/[^{12}C])\ \text{sample}}{([^{13}C]/[^{12}C])\ \text{standard}} - 1] \cdot$$

The standard can be taken to be a source of inorganic carbon, available to living organisms, as the marine carbonate. Higher plants, algae, photosyn-thetic bacteria and chemosynthetic archaebacteria show $\delta^{13}C$ clustered around $\delta^{13}C \simeq {,}0.025$. Similar analysis have been performed on ancient sediments and rocks of putative biological origin with the impressive result that a similar isotopic fractionation is found all the way back to about 3.5×10^9 years ago. This result not only reinforces the interpretation of apparent findings of microfossils, but also suggests that the basic biochemical mechanisms of carbon assimilation have remained remarkably similar through all of biological evolution.

Finally much effort has been spent to identify molecular fossils, that is organic molecules originally synthesized by ancient living systems. Such molecules have obviously been modified while residing in sediments and rocks. However it has been possible to assign with some confidence ages of more than 10^9 years to aminoacids, porphyrins, fatty acids and aromatic and saturated hydrocarbons.

Whatever the process and the ambient which stimulated chemical evolution on the Earth after its formation, it must have been very fast, because such a chemical evolution gave rise, apparently within as little as 10^9 years, to complex multimolecular structures, which were compartmentalized, could metabolize carbon products reproduce and duplicate.

Can we advance plausible pathways for such a chemical evolution from the knowledge we have of the Earth and the nearby interstellar medium and from chemical processes we can reproduce in the laboratory? The answer is largely yes for virtually every oligomeric organic molecule of biological interest as the building blocks of proteins, nucleic acids and membranes and prosthetic groups and cofactors. The situation is less happy with more complex biological molecules and structures, while the gap between such structures and any hypothetical protocell is completely unfilled. Let us consider the very early chemical evolution, the so called prebiotic evolution. There are a few "scenarios" which are currently popular.

The first invokes the classical picture of the hot primordial Earth. It is assumed that the atmosphere of the early Earth was rich in hydrogen, methane, ammonia and water; and that chemical reactions were initiated by high temperatures, say hundreds of °C, electrical discharges, ultraviolet light, and α, β and γ rays. The pioneering experiment of Miller and Urey showed that urea and various aminoacids were produced by spark discharges in a gaseous mixture of CH_4, NH_3, H_2 and H_2O. Since then many experiments have been performed in solid, aqueous and gaseous phases. These have started with a variety of initial chemical compositions, including the use of some of the products obtained in the Miller experiment, such as CO, HCN, formaldehyde etc., and have made use of many different energy sources. The result has been the synthesis of virtually all of the small organic molecules relevant to biochemistry, including the sugars ribose and deoxyribose, adenine, simple fatty acids, aminoacids, etc.. Then, starting from such molecules and in similarly hot conditions, it has been possible to obtain evidence of the formation of more complex molecular structures with greater biological relevance such as AMP, ADP, ATP, nucleotides, nucleoside triphosphates, and porphyrins. Finally from mixtures of aminoacids, thermally treated, Fox has been able to produce chains of aminoacids with molecular weights from 4 000 to 10 000 Daltons. Such "proteinoids" exhibit many properties in common with contemporary proteins including a number of "enzyme-like" activities. Therefore, protein-like polypeptide chains, with rudimentary biocatalytic properties, can be produced by non-specific, non-instructed chemical pathways. Not so for polynucleotides. Similar attempts at "spontaneous" synthesis of such structures under energetically favorable, hot but otherwise non-specific conditions, fail to produce polynucleotide chains more than five bases in length.

More recently another scenario has started to attract considerable interest. The primordial Earth is postulated to have been "contaminated" from space with significant amounts of biologically relevant small molecules, or even, in the most daring speculations, with complex organized prebiotic structures. Then the favorable conditions found at some locations on our planet could have been exploited for the subsequent evolution to living systems. An interstellar chemistry was inconceivable a few decades ago. If chemical kinetics would simply obey the Arrhenius law, which states that kinetic constants vary as $e^{-\Delta E/k_B T}$, where ΔE is an activation energy intrinsic to the chemical reaction under consideration, then outer space at $\sim 3\,°K$ is too cold. For any reaction leading to the production of even the simplest molecules from interstellar atoms and ions, the rate would be negligibly small and one should not expect to find any significant amount of matter in molecular form in the interstellar medium. Various possibilities have been suggested to account for the observed enhancement in the synthesis of complex molecules. As atoms in the interstellar space are constantly ionized by high energy particles an ion-molecular chemistry can take place, that is chemical reactions between ionized species, which occur much faster than the reactions between neutral atoms. Another possibility is that grains of interstellar dust may provide catalytic effects on their surface, again to enhance rates of chemical reaction. Finally, instead of overcoming the energy barrier ΔE by thermal activation, the atoms or ions involved may tunnel through the barrier by the well known quantum mechanical process. Such molecular chemical tunneling would make the reaction rates temperature independent at low temperature, where this process would be dominant in respect to thermal activation. Laboratory experiments indicate, for the relevant case of the polymerization of formaldehyde, that such a process occurs at temperature below $\sim 10\,°K$, making the kinetic rates many orders of magnitude greater than estimated by extrapolating the Arrhenius plot of rates at higher temperatures. Spectroscopic observations of the interstellar medium in the radio frequency range have recently shown lines typical of the rotational spectra of molecules of a variety of atomic compositions and complexities. The intensity ratios of these lines, when a Boltzmann distribution is assumed among the energy levels, also gives the temperature of the medium where the molecules are found. By this method giant molecular clouds have been discovered. Some are relatively warm, 20 to $100\,°K$; dense, 10^6 to 10^7 molecules/cm^3 and massive, $10^3 \div 10^5$ solar masses, such as the Orion Molecular Cloud. Others are relatively cold, $10 \div 15\,°K$ and less dense,

$10^4 \div 10^5$ molecules/cm^3, such as the Taurus Molecular Cloud. Both types of clouds are sites of the formation of newborn stars, by the way. Table 7-III gives an impressive list of inorganic and organic molecules of various complexity found in the Orion and Taurus clouds.

TABLE 7-III

Some complex molecules identified in dense molecular clouds

hydrides, oxides, sulfides	H_2, OH, H_2O, CH, CS, CO NH_3, N_2H^+
nitriles, acetylene derivatives	CN, HCN, CH_3CN, C_2H_5CN, CH_3C_2H, CH_3NH_2, HC_3N, CH_3CH_2CN, HC_5N, HNCO,
aldehydes, alcohols, ethers	H_2CO, $HCOOCH_3$, $(CH_3)_2O$, H_2CS, CH_3OH

Similarly, the spectroscopic examination of comets in the visible, ultraviolet, infrared and radio frequency regions has uncovered the presence of a variety of molecular species in these bodies. At present it is believed that significant quantities of cometary material have been and still are being accreted by the Earth as it moves in the interplanetary medium. In addition the possibility of the direct accretion of interstellar particles when the solar system passes through a dense interstellar molecular cloud should not be overlooked. Therefore, there are clearly mechanisms by which biologically relevant molecules could have transferred from space to the Earth. Finally the extent to which chemical processes have proceeded in space, leading to the production of organic molecules, is testified to by the finding in the Murchison meteorite of 17 different aminoacids, 7 of which are among the 20 kinds normally found in the proteins of living organisms.

So, either by synthesis in its own "warm little ponds", or by accretion from space, or both, the early Earth could have acquired suitable amounts of the substances which we consider to be basic for living systems, either as precursors or as actual biological molecules. The next problem is how did these molecules become organized into a structure sufficiently complex to serve as a prototype of an actual living system.

By some sort of "spontaneous" self-assembly, we need to produce self-replicating molecules, the properties of which could induce the organization of matter into a living system and eventually to evolve a molecular mechanism for the storage of the information for that organization. How to produce a reasonably plausible pathway for such a process is in practice not very clear. Unless highly specialized enzymes are present, there is no synthesis of long polynucleotides chains from activated nucleotides. Possibly this gap will be bridged sometime soon. For the time being let us examine what can happen to a population of such self-replicating molecules, assumed to be already existing in an aqueous solution, where their energy-rich precursors are also present. The discussion of such a model is of great interest, because it permits one to reduce certain aspects of the basic law of biology, Darwin's survival of the fittest, to a "theorem" in the physical chemistry of molecules in solution. The attempt is to produce a model of molecular evolution which through self-replication, mutability and metabolism, will display competition and selection among a variety of molecular species. Such a model would reproduce the essential features of a Darwinian system on one side and on the other side it could also serve the purpose of bridging the gap between the primordial chemical evolution leading to prebiotic molecules and the appearance of organized living structures, after which Darwinian biological evolution could satisfactorily account for the observed patterns.

Following Eigen, we can translate the distinctive features of living systems, self-replication, mutability and metabolism into molecular properties, for which an operational physical definition can be formulated. This will permit us to write a dynamical equation for the time evolution of the concentration of each molecular species. It will be interesting to see that the different molecular species will be in competition for the smaller "building block" molecules and will evolve taking advantage of mutations which result in larger and larger values of a parameter, the selective value. All this will be due to the intrinsic non-linearity of these equations, which in turn is due to the fact that the system is in a steady state far from equilibrium. The result is an attractive model, which has recently been shown to operate in actual experimental situations.

To get what we would call self-reproduction at the molecular level, we use the notion of autocatalysis: the ith molecular species participates, with catalytic properties, in the synthesis of copies of itself from smaller molecules, so that its concentration x_i will change in time according to

$$\frac{dx_i}{dt} = A_i Q_i x_i - D x_i + \Sigma_{k \neq i} W_{ki} x_k + \phi_i \tag{1}$$

where A_i is a rate parameter which quantitates the autocatalytic properties of i, and Q_i is a quality factor which quantitates the fidelity of reproduction, $0 < Q_i < 1$. Out of the $A_i x_i$ copies produced by autocatalysis, only $A_i Q_i x_i$ are identical to i. While A_i may be a function of the concentrations $\{m_\lambda\}$ of the set $\{\lambda\}$ of "monomers" which, steadily supplied to the solution, constitute the "building blocks" to synthesize the species $\{i\}$, Q_i may be an intrinsic property of the molecular recognition process. The species are removed from the solution either because of spontaneous decomposition, as described by the term Dx_i, or because the specie i may diffuse or be transported out of the system. While the $A_i(1 - Q_i)x_i$ imprecise copies of the specie i result in the increase in population of other species $k \neq i$, error copies from $k \neq i$ may increase the population of i, as described by the term $\underset{k \neq i}{\Sigma} W_{ki} x_k$.

This model shows "self-reproduction", because of the autocatalytic term $A_i Q_i x_i$, "metabolism" because of the "spontaneous" formation and decomposition terms and of the flux term ϕ_i and "mutability" because of the quality factor Q_i and the cross terms W_{ki}. The parameter $W_{ii} = A_i Q_i - D_i$ has been given the name of "selective value". For a specie to survive and win the competition for appropriating the oligomers, not only must it display $W_{ii} > 0$ but also W_{ii} must be larger than a threshold value, which depends on the selective values of every other specie and on the external constraints. In this way the survival of the fittest in a given environment is given a meaning in terms of molecular properties.

To see this in more detail let us look into a solution of the dynamical equations for such a system under convenient constraints. We assume that the set of species $\{i\}$ exhausts all the species present in solution. Therefore any error copy from misreproduction of i goes into the production of k and obviously

$$\Sigma_i A_i (1 - Q_i) x_i = \Sigma_i \underset{k \neq i}{\Sigma} W_{ki} x_k . \tag{2}$$

Then we assume that the overall level of organization is constant in time. To do so we write an outflow ϕ_i of the products, proportional to the concentrations x_i

$$\phi_i = \phi_0 \frac{x_i}{\Sigma_k x_k} \tag{3}$$

to remove the total excess production of organized material, so that

$$\Sigma_k (A_k - D_k) x_k = -\phi_0 \tag{4}$$

and we keep the average of concentration of all species constant

$$\Sigma_k x_k = C . \tag{5}$$

We also buffer the oligomeric building blocks at constant concentrations. This means that, as A_i depends on these concentrations and we now disregard any dependence on the x_k, we have that the term $A_i Q_i$ is a constant. With these assumptions we can write the dynamical equation of evolution as

$$\frac{dx_i}{dt} = (W_{ii} - \overline{E}(t))x_i + \sum_{k \neq i} W_{ki} x_n \tag{6}$$

where

$$\overline{E}(t) = \frac{\Sigma_k (A_k - D_k) x_k}{\Sigma_k x_k} = -\frac{\phi_0(t)}{C} \tag{7}$$

is the average excess productivity removed by the flow term $\phi_0(t)$.

This formulation of the rate equation clearly shows the intrinsic non linearity of the process, because of the term $\overline{E}(t)$. The competition between species arises because each has to achieve $W_{ii} > \overline{E}(t)$ in order to survive, but $\overline{E}(t)$ depends on the concentration of all the others. The form of $\overline{E}(t)$, which is left invariant under a linear transformation $x_i \rightarrow y_i$, permits one to rewrite the equation in terms of the eigenvectors y_i and the eigenvalues ψ_i of the associated linear equation as

$$y_i = (\psi_i - \overline{E}(t)) y_i \tag{8}$$

where

$$y_i = \Sigma_k a_{ki} x_k , \qquad \Sigma_i y_i = \Sigma_i x_i \tag{9}$$

and

$$\overline{E}(t) = \frac{\Sigma_k \psi_k y_k}{\Sigma_k y_k} \tag{10}$$

This form of the equation clearly shows what we have been anticipating above. $\overline{E}(t)$ increases with time since any y_i with a relatively larger ψ_i will tend to dominate. Therefore the threshold condition for survival, that is $\psi_i > \overline{E}(t)$, is continuously pushed higher until it reaches a steady state for $\overline{E}(t) \rightarrow \psi_{max}$. Eigen has called a "quasi-specie" the population y_i which arises from a linear combination of the x_i. The quasi-specie will contain, in actual numerical solutions, a leader sequence and other sequences, which closely resemble the leader, as they have very few errors. They coexist with the leader, as their selection values are very close to that of the leader. This behavior is similar to that of neutral mutations in actual living systems, when a mutation, which is not particularly harmful nor advantageous, is propagated in the population together with the leader genotype.

As the eigenvalues ψ_i are expressed in terms of the W_{ii} and W_{ik} through the linear transformation, we see the roots of the selection process in the molecular properties of system. Further insight into such a relationship is given by considering a perturbative approximation to calculate explicity ψ_{max}. This gives, to second order,

$$\psi_{max} \simeq W_{mm} + \Sigma_{k \neq m} \frac{W_{mk} W_{km}}{W_{mm} - W_{kk}} \tag{11}$$

where W_{mm} is the selective value of the dominant specie. This is valid if none of the W_{kk} for $k \neq m$ approaches W_{mm} too closely. Now we manipulate $\overline{E}(t)$ to extract the average productivity of all the species other than the dominant one. We get obviously

$$\overline{E}(t) = \overline{E}_{k \neq m}(t) + \frac{x_m}{\Sigma_k x_k}(E_m - \overline{E}_{k \neq m}) \tag{12}$$

where

$$\overline{E}_{k \neq m}(t) = \frac{\underset{k \neq m}{\Sigma} E_k x_k}{\underset{k \neq m}{\Sigma} x_k} \tag{13}$$

If we now neglect in the equation for ψ_{max} the summation term, which represents the backflow from other species, we can rewrite the steady state condition $\psi_{max} = \overline{E}(t)$ to give the steady state concentration $\overline{x}_m$ of the dominant specie

$$\frac{\overline{x}_m}{\Sigma_k x_k} = \frac{W_{mm} - \overline{E}_{k \neq m}}{E_m - \overline{E}_{k \neq m}} \tag{14}$$

Therefore the condition for successful selection becomes

$$W_{mm} > \overline{E}_{k \neq m} . \tag{15}$$

The dominant specie is that for which the selective value is larger than the average productivity of all other species present. We may recall that we have obtained this result under a specific set of boundary conditions and in an approximation where second order terms have been neglected. Still, as we shall see shortly, these conditions can be used in actual laboratory experiments, which give results in agreement with the theory.

Before we do so let us further elaborate on the selection condition above to relate it to the frequency of copying errors and to the maximum number of symbols in the informational molecular species which still permits stability of the information content. As $W_{mm} = A_m Q_m - D_m$ we find that, in order to satisfy the condition for successful selection, eq. (15), we must have that

$$Q_m > Q_{\min} = \frac{D_m + \overline{E}_{k \neq m}}{A_m} \tag{16}$$

In general the quality factor Q_i in replicating the molecular specie i of ν_i symbols can be expressed in terms of the quality factors $q_{\alpha i}$ for the correct reproduction of each symbol α in the molecule of the specie i. We have obviously

$$Q_i = \Pi_\alpha^{\nu_i} q_{\alpha i} = <q_i>^{\nu_i} \tag{17}$$

if we take the average fidelity $<q_i>$ as the average of the $q_{\alpha i}$. If for a dominant specie m, with a specific sequence of ν_m symbols, eq. (16) above is satisfied then at steady state the concentration of this specie will remain constant despite ongoing synthesis and decomposition. Information will be conserved and such a steady state will be stable. The condition eq. (16) for Q_m means that there is a maximum length of the "message", i.e. a maximum number ν_m of symbols in the gene which can be conservatively replicated. This limitation in ν_m is given by

$$\nu_m < \frac{\log \left(\dfrac{A_m}{D_m + \bar{E}_{k \neq m}} \right)}{1 - <q_m>} \qquad (18)$$

where we have approximated $\log <q_m> = \log [1 + <q_m> - 1] \simeq <q_m> - 1$, as $1 - <q_m> \ll 1$. Again we can connect a global property of the replicating dominant sequence with its microscopic physico-chemical properties.

RNA replication, assisted by a specialized replicase enzyme, seems to allow $\nu_{max} \simeq 10^3$ to 10^4. This is suggested by the fact that none of the single stranded RNA phages has more than 10^4 bases in its genome and is in agreement with experimental studies on the molecular properties which give independent estimates of $<q_m>$ and of Q_{min} of $1 - <q_m> \simeq 5 \cdot 10^{-4}$ and $Q_{min} \simeq 10^2$. The replication of DNA in bacteria, assisted by polymerases and proofread by exonucleases, would have $1 - <q_m> \simeq 10^{-6}$ and $Q_{min} \simeq 10^2$, which would permit $\nu_{max} \simeq 0.5 \div 5 \cdot 10^6$, which compares well the genome of E. coli, which possesses 4×10^6 bases. Non-enzymatic replication of nucleic acids has never been observed to any significant extent and the molecular properties suggest that at most $<q_m> \simeq 0.90 \div 0.99$ so that for $Q_{min} \simeq 10^2$, one would expect $\nu_{max} \simeq 100$. This might be compared with the number of bases in tRNA, $\nu \simeq 80$, and it has suggested to Eigen that tRNA might be the "oldest" replicative unit. However a reasonable process which would catalyze the synthesis ex novo and the replication of RNA strands of length $\nu > 5$ in absence of specific enzymes has never been produced in the laboratory.

The theory outlined above has been successfully tested in laboratory experiments in vitro. As the primary self-replicating molecule these experiments used the RNA of a virus which infects E. coli, the $Q\beta$ phage. It has a genome of 4 500 bases, which in part codes for a specific enzyme, the $Q\beta$ replicase, needed to assist replication of the RNA strand. It was shown by Spiegelmann that synthesis of infectious viral RNA can be produced in vitro starting with activated nucleotides ATP, GTP, UTP and CTP, the replicase enzyme and even as little as a single copy of the viral RNA.

In order to promote "evolution" in vitro of the RNA sequence, a mixture is incubated for a predetermined period and then a small fraction of the products are transferred to fresh solutions, which have the same enzyme and precursor concentrations as the initial one but no other RNA copies. This serial transfer method enables one to vary considerably the evolutionary constraints on the system. Results display many of the features predicted by the model above. For instance the sequences of the RNA copies, produced from a single template in the molecular competition for the monomers, are not all completely identical even though they remain highly infectious. Only a small fraction, of the order of a few percent, is exactly identical to the initial sequence, while the rest resemble it very closely, with only one or a few errors. This is the notion of a quasi-specie.

In other experiments one started with a mutant prepared in vitro which has a single base error in a specific location and observed its evolution in competition with the "wild type" or standard sequence. This experiment allows a quantitative evaluation of the distribution in mutant and wild type populations and of the difference in selective values of these two sequences. The results indicate a substantial agreement with the selective values calculated from the molecular parameters determinated separately.

Striking results are obtained if one promotes fast self-reproduction by allowing shorter intervals between subsequent serial transfers. Under these new selection constraints, the RNA quickly evolves. Infectivity is lost after the fourth transfer in the series and the dominant sequence becomes shorter and shorter until after 76 transfers a constant sequence of some 550 nucleotides is retained. Still shorter intervals lead to even shorter sequences of some 200 nucleotides. The $Q\beta$ replicase enzymes recognizes the RNA template through only part of its sequence and shorter copies are produced in any condition. When short intervals are imposed between transfers, these copies acquire under the new constraints a higher selective value and therefore tend to dominate over the longer sequences. As the constraint of infectivity is also removed the dominance of any shorter, non-infectious, species becomes overwhelming.

Finally a quite striking experiment has been performed recently by this method of serial transfer. Activated monomers and highly purified $Q\beta$ replicase were incubated without any RNA template. In these conditions, after incubation times of a few hours, in contrast to incubation times of the order of tens of minutes when templates were present, synthetic polynucleotide chains similar to the Spiegelmann's minivariant were observed. These experiments of synthesis without template produce products which are uniform when the system is not disturbed during the incubation time. If on the contrary, in the middle of the incubation time, one splits the solution in various compartments, still keeping in each the optimal conditions for synthesis without template, the synthesis gives products which are uniform within each compartment, but different in different compartments. From this it seems that the dominating sequence is the result of chance initial conditions at the critical size of a polymer of a few nucleotides and that no instructions actually come from the enzyme, in agreement with the central dogma.

This last experiment bears some interesting significance on the origin of the self-reproductive molecule. We have noted that so far experiments have failed to produce such a molecule without specific catalysis, as it was possible for proteinoids. However, here we see a possibility of bridging this gap and completing the picture of the prebiotic evolution. This view may be a bit too optimistic at this stage, since the $Q\beta$ replicase is a highly specific enzyme and it is not clear how and if some of the proteinoids, which may have enzymatic activity could have taken the same role in early evolutionary stages. However, one cannot help wondering if some of the spontaneously formed proteinoids might have had the catalytic activity necessary for the primordial, non-instructed synthesis of polynucleotides.

The model of molecular evolution outlined above, cannot account for the emergence of the more complex compartmentalized self-replicating structures, which could have served as primordial protocells. Indeed we have seen that: i) a molecule with autocatalytic properties under optimal conditions, kept constant by the external world in a sort of "biological paradise", shows the features of a Darwinian like molecular evolution; ii) the conditions of selective pressure needed to produce such an evolution are directly linked to physical properties of the constraints and of the molecule itself, so that in this sense selection acts directly on the "genome"; iii) the outcome of the evolution is not affecting the overall conditions in which it is taking place. In contrast, to bridge the gap between any prebiotic molecular revolution and the primordial protocell population, we would like to see: how a self-replicating, informative molecule, the nucleic acids, did get linked to a system of operative molecules, the proteins; how in the process an (almost) universal genetic code evolved; how because of this linkage the fidelity of replication increased in such a way as to allow genomes as long as over 10^9 bases as seen now; how the self-replication did involve at some point compartmentalized, complex structures; and how the selective pressure which now acts on such individuals as a whole is related to the continuing evolution of the informative molecule. These problems are enormous and research in a variety of their many facets is quite lively at present. We give here only a very brief mention of recent developments and refer for further reading to the bibliography at the end of the chapter.

Eigen and Schuster have proposed the "hypercycle" as the intermediate stage between a population of, say, RNA-like molecules, which would replicate without specific enzymes with an accuracy to permit about 10^2 bases, and a population of compartmentalized structures, the RNA of which codes for a specific enzyme to give an accuracy consistent in turn with the 10^4 bases needed

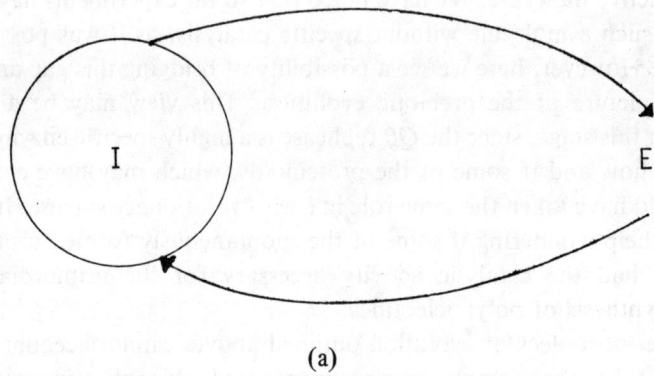

(a)

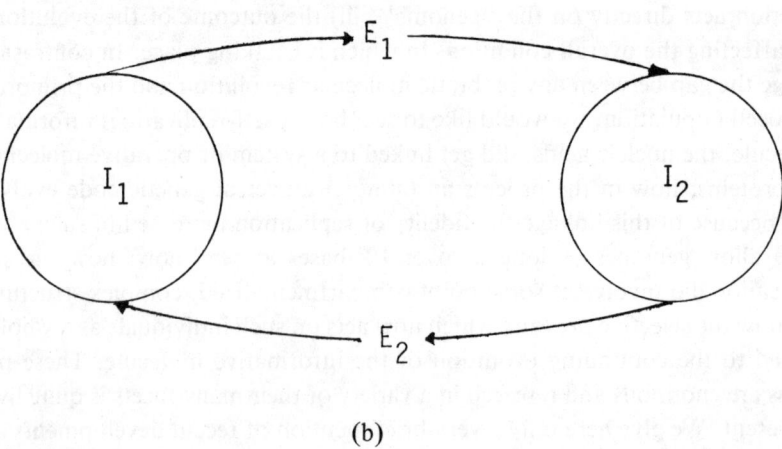

(b)

Fig. 7-2 Simplest hypercycles: a) the "informative" component I
instructs the synthesis of the "operative" component E, which assists
I in its self-replication b) I_1 codes for E_1, which assists I_2 in its
self-replication, which codes for E_2, which assists I_1 in its self-
replication.

to code for a primitive protein-synthesizing machinery. A simple example of hypercyclic linkage between self-replicating molecules and catalysts is given in Fig. 7-2. The self-replication of I is assisted by the catalyst E, which in turn is produced by a process of synthesis instructed by I. A RNA phage works just similarly by using the host translation machinery to produce E from the instruction coded in I. Notice that now, in order to recognize specifically I among the population of host RNA strands, the enzyme E has to rely on the molecular structure and shape taken up by I in the environment, that is to say that E must rely on phenotypic properties of I, which, although obviously related to the genotypic properties of I, however do not come in a one to one correspondence, as they depend also on the solution conditions of the environment. Eigen and Schuster have made extensive theoretical studies of the properties of hypercycles of various complexity and in a variety of situations of interest. Some of the general properties they get are the following: i) the set of all the molecular species connected with the hypercyclic linkage coexists in a stable and controlled manner, grows coherently, competes with any replicative unit not belonging to the hypercycle and may enlarge or reduce its size if selective advantage is offered, ii) the internal linkages and cooperativities evolve towards optimal, with a process of direct fixation for those of the phenotypic kind, which give advantage directly to the reproductive properties of the mutant, while for those of the genotypic kind, which would favor a subsequent product and hence only indirectly the mutant, compartmentalization it is required before competitive fixation may occur, iii) the hypercycles are not Darwinian systems: while in a Darwinian system the advantageous mutant can establish itself irrespective of its initial population size, the selective advantage for hypercycles is intimately linked to their population size and so mutants, as they always emerge as few copies, cannot easily replace the wild type. The last property says that the selection of a specific hypercycle among others is a "once-for-ever" decision. This appears to be consistent with the universality of the genetic code: if the code got established through a process of evolution via competition, then the hypercycle model could explain its universality. It is also interesting to see how the requirement of compartmentalization is directly related to the possibility of evolving a larger genome, with its more accurate reproduction mechanisms. With the notion of the hypercycle as the "missing link", Eigen and Schuster developed a proposal for the physics of primordial coding and of the early translation apparatus, which among other merits, concludes that the limited number of primordial codes was GGC, GCC, GAC, GUC in nice agreement with the fact that in simulated prebiotic synthesis of aminoacids by far the most abundant products are glycine and alamine, presently coded by GGC and GCC, and then aspartic acid and valine, coded now by GAC and GUC respectively.

Are there any traces of such an evolution of the genetic code? It has been quite recently discovered that there are exceptions to the universality of the genetic code. In mitochondria from yeast and mammals there are deviations from the standard code. The triplet UGA codes for tryptophan instead of acting as a termination signal. In yeast codons of the type CUN, where N is any of the four bases, all code for threonine instead of leucine as in the case of the standard. For codons of the type AUN, the deviations appear only for mammals. So not only do mitochondria show deviations, although limited, from the standard code, but also there is a variability between mitochondria of different species. If, as it has been proposed, one regards mitochondria as primordial bacteria, which after prolonged symbiosis have been stably incorporated in the cells of higher organisms, then one is allowed to regard these deviations as the few remnant traces of the evolution of the code, as the incorporation in the cells of different species put different codes in biologically protected niches, where they could survive against the dominance of the leader code.

Finally we would like to discuss briefly another major problem in the origin of life, the origin of handedness of the molecules involved in the living systems. Many molecules of biological relevance as building blocks, metabolites, etc. can exist in two enantiomers of intrinsic left-handed and right-handed spatial symmetry respectively. Fig. 7-3 shows the aminoacid alanine in its two possible enantiomeric forms: L-alanine is a mirror image of D-alanine and, despite the identity of the two in composition and chemical bonds, they are not superimposable one on the other by any continuous spatial symmetry operation. Of course the same applies to all aminoacids. Non biological synthesis of aminoacids in close to equilibrium conditions leads to products which are half and half mixtures of the two enantiomers. Attempts to use left-right symmetry breaking fields to induce in the products a relative abundance of one enantiomer at the expense of the other, as for instance combinations of electric, magnetic and gravitational fields, have shown that the discrimination achievable would be vanishingly small for any realistically attainable field strength. Therefore one wonders why, for instance, all living systems use only left-handed aminoacids and right-handed sugars. A classic answer to this question is that it has been a matter of chance. As soon as the biochemistry of the primordial living systems became compatible with only one type of enantiomer, whichever the choice, then by chance the left-handed aminoacids were picked up and the choice got stabilized in all subsequent generations. We would not be able to digest the meat of an hypothetical animal whose proteins would be of right-handed aminoacids. By making chance intervene, the symmetry breaking choice made by living systems is reconciled with the requirement that the physical laws of nature should be indifferent to an overall change of left into right, that is that there is no experiment that can decide if we are in an intrin-

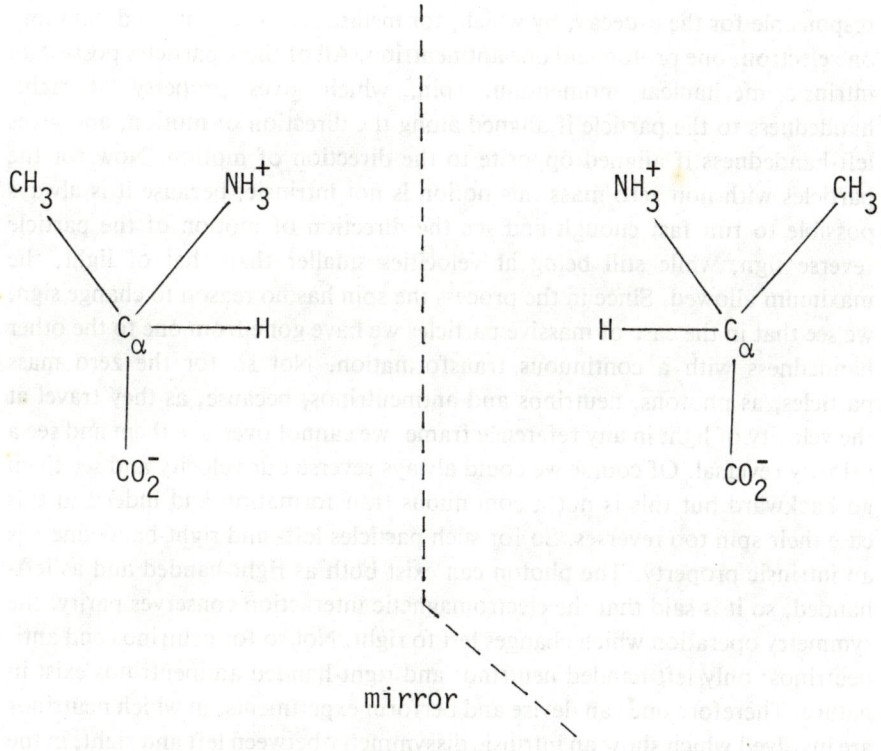

mirror

Fig. 7-3 The aminoacid alanine in its two enantiomeric forms, the left handed L-alanine (left) and the right handed D-alanine (right); one is the mirror image of the other. The hydrogen bond connecting the hydrogen atom to the central alpha carbon sticks out of the plane of the paper: when in going from the CO_2 group to the NH_3^+ group one finds the CH_3 residue respectively on the left in L-alanine and on the right in D-alanine.

sically left- or right-handed world. If a macroscopic discrimination is found, it has been a matter of chance in the initial conditions. In another similar instance the choice may well be the opposite; and it would in a statistics of many trials, as in growing crystals of enantiomorphous molecules, which come out half left- and right-handed respectively.

However an intrinsic left-right dissymmetry of nature has been somewhat unexpectedly discovered a few years ago. It has been found in the realm of the fundamental interactions which govern the subatomic world. The so called "weak interaction" is said to be parity violating. The weak interaction is responsible for the β-decay, by which, for instance, a free neutron decays into one electron, one proton and one antineutrino. All of these particles possess an intrinsic mechanical momentum, spin, which gives property of right-handedness to the particle if aligned along the direction of motion, and gives left-handedness if aligned opposite to the direction of motion. Now for the particles with non zero mass this notion is not intrinsic, because it is always possible to run fast enough and see the direction of motion of the particle reverse sign, while still being at velocities smaller than that of light, the maximum allowed. Since in the process the spin has no reason to change sign, we see that in the case of massive particles we have gone from one to the other handedness with a continuous transformation. Not so for the zero mass particles, as photons, neutrinos and antineutrinos, because, as they travel at the velocity of light in any reference frame, we cannot overtake them and see a velocity reversal. Of course we could always reverse our velocity and see them go backward but this is not a continuous transformation and indeed in this case their spin too reverses. So for such particles left- and right-handedness is an intrinsic property. The photon can exist both as right-handed and as left-handed, so it is said that the electromagnetic interaction conserves parity, the symmetry operation which changes left to right. Not so for neutrinos and antineutrinos: only left-handed neutrinos and right-handed antineutrinos exist in nature. Therefore one can devise and perform experiments, in which neutrinos are involved which show an intrinsic dissymmetry between left and right, in the sense that the outcome of the "mirror" experiment simply does not occur in nature. This happens only for the weak interaction.

Can this intrinsic left-right dissymmetry of one of the fundamental interactions help any in understanding the left-right dissymmetry of biomolecules? At first sight an affirmative answer may appear a bit far fetched, because one may expect the effect of the weak interaction to be limited to the subatomic world. Quite recently however the view has emerged that such a connection is not unreasonable. According to theories and experiments of the last few years, the weak and electromagnetic interactions are aspects of a single "electroweak" force. As a consequence the parity violating properties of the weak interaction would perturb atomic and molecular energy levels by tiny amounts, which are however different for left- and right-handed molecules. It has been calculated that the energy difference in the ground state between L-alanine and D-alanine is about $6 \times 10^{\pm 19}$ eV, the left-handed L-alanine being more stable by that amount. Such a small difference would not be very important in close to equilibrium processes, as it would give rise only to an excess of some 10^6 L-aminoacids in one mole of a mixture. But in an open, far from equilibrium system the situation could be different. Schemes involving autocatalysis have been proposed according to which the synthesis of molecules with intrinsic left- or right-handedness from molecules without such property can give products in which one of the two enantiomers has been discriminated in respect to the other. This would happen only at special values of parameters, as the reactant concentrations, which characterize the far from equilibrium process, where the system becomes hypersensitive to very small energy differences between the two enantiomers. So, according to this view, the parity breaking effects in living systems are not a result of chance, but are a compulsory outcome, which ultimately is linked to very basic laws of nature.

FURTHER READINGS

General

Fox, S. W. and Dose, K. (1977) *Molecular evolution and the origin of life*, M Dekker Inc.

Ponnampemura, C. ed., "Cosmochemistry and the origin of life", (1983) *Proc. NATO Adv. Study Inst. Maratea, Italy July 1981*, D. Riedel Publ. Co.

Hoyle, F. and Wickramasinghe, C. (1978) *Lifecloud*, J. M. Dent and Sons.

Eigen, M. and Schuster, P. (1978) *The hypercycle. A principle of natural self-organization*, Springer-Verlag.

Specific

Sagan, C. S. and Drake, F. "The search for extraterrestrial intelligence", (May 1975) *Sci. Am.* **232**, 80.

Papagiannis, M. D. "Recent progress and future plans on the search for extra-terrestrial intelligence", (1985) *Nature* **318**, 135.

Barghoorn, E. S. "The oldest fossils", (May 1971) *Sci. Am.* **224**, 30.

Groves, D. I., Dunlop, J. S. R. and Buick, R. "An early habitat of life", (Oct. 1981) *Sci. Am.* **245**, 56.

Schopf, J. W. "The evolution of the earliest cells", (Sept. 1978) *Sci. Am.* **239**, 84.

Dickerson, R. F. "Chemical evolution and the origin of life", (Sept. 1978) *Sci. Am.* **239**, 62.

Goldanskii, V. I. "Facts and hypotheses of molecular chemical tunnelling", (1979) *Nature* **279**, 109.

Schuster, P. "Prebiotic evolution", Ch. 2 in *Biochemical evolution*, Gutfreund, H. ed., Cambridge University Press (1981).

Eigen, M., Gardiner, W., Schuster, P. and Winkler-Oswatitsch, R. "The origin of genetic information", (April 1981) *Sci. Am.* **244**, 78.

Eigen, M. "How does information originate? Principles of biological selforganization", in Rice, S. ed., (1978) *Adv. Chem. Phys.* vol. XXXVIII, J. Wiley.

Margulis, L. "Symbiosis and evolution", (Aug. 1971) *Sci. Am.* **225**, 48.

Mason, S. F. "Origins of biomolecular handedness", (1984) *Nature* **311**, 19.

INDEX